Poster Sessions

Held at the 2023 AIChE Annual Meeting

Orlando, Florida, USA
5-10 November 2023

ISBN: 978-1-7138-9321-9

Printed from e-media with permission by:

Curran Associates, Inc.
57 Morehouse Lane
Red Hook, NY 12571

AIChE
120 Wall Street, FL 23
New York, NY 10005-4020

Phone: (800) 242-4363
Fax: (203) 775-5177

www.aiche.org

Additional copies of this publication are available from:

Curran Associates, Inc.
57 Morehouse Lane
Red Hook, NY 12571 USA
Phone: 845-758-0400
Fax: 845-758-2633
Email: curran@proceedings.com
Web: www.proceedings.com

TABLE OF CONTENTS

GENERAL POSTER SESSION

84a Machine Learning with Weighted-Soap to Efficiently Predict Electron Densities

Siddarth Achar, Leonardo Bernasconi, Karl Johnson

We address the computational complexity limitations of density functional theory (DFT) in predicting electron densities for materials chemistry applications. We have developed a deep learning formalism, the Deep-learning Charge Density Prediction (DeepCDP) method, which can be used to generate 3D charge density maps using only atomic positions as input data. Our method is trained on DFT electron densities for small systems and can predict densities for arbitrarily large systems. We used weighted-Smooth Overlap of Atomic Orbitals (w-SOAP) descriptors to fingerprint atomic environments and map this information onto their electron densities on a grid-point basis. Our DeepCDP formalism was tested for both molecular and periodic systems, achieving electron density prediction accuracies in excess of 99% for large system sizes (>900 atoms) using models trained on small systems (≤10 atoms). DeepCDP also achieves similar accuracies for both charged and uncharged systems, making it possible to address phenomena like proton diffusion on large sheets of functionalized graphene over time scales sufficient to estimate proton diffusion barriers with DFT accuracy. Additionally, the ability of DeepCDP's to exploit the locality of the electron-electron interactions allows us to bypass the memory inefficiencies of common DFT implementation schemes, resulting in rapid and accurate electron density predictions for arbitrarily large systems. This development has significant implications for materials chemistry research and paves the way for further advancements in computational chemistry.

CONFINED FLUID PHASE BEHAVIOR OF CARBON DIOXIDE IN NANOPOROUS MEDIA

Omer Salim [*], *Keerti Sharma, and Mohammad Piri*

Center of Innovation for Flow through Porous Media, Department of Energy & Petroleum Engineering, University of Wyoming, Laramie, WY 82071, USA

Abstract

Geological formations possessing nanometer-sized pores, such as shale reservoirs, are promising candidates for sequestering vast quantities of carbon dioxide. To improve the current understanding of the storage mechanisms in such subsurface systems, it is essential to conduct further research on the fundamental physics of storage and phase behavior of CO_2 confined in nanopores. This work aims to answer some critical questions vis-à-vis the thermodynamic properties of confined subcritical CO_2. To this end, a systemic experimental investigation was performed utilizing the model adsorbents, a silica-based nanoporous medium (MCM-41), at varying pore diameters and temperatures. We employed a novel gravimetric apparatus to experimentally measure the adsorption and desorption isotherms and study the phenomena of capillary condensation and evaporation of CO_2. MCM-41 samples of two different pore sizes (6 and 10 nm) were used as the adsorbent, and the isotherms were generated at temperatures ranging from -23.1 to 20 °C. The onset and end pressures of the capillary condensation and evaporation processes and the capillary condensation mean pressure were determined for each pore size. The experimentally measured bulk condensation pressures were successfully compared against their NIST-reported counterparts. The results reveal that the vapor-liquid phase transition pressure of CO_2 increases with decreasing temperature for the same pore size. Similarly, capillary adsorption increases as the pore size is reduced at a fixed temperature, confirming the predominance of the confinement effect in pores with smaller diameters. The results show that, due to confinement, CO_2 can be stored as a liquid phase inside the nanopores at much lower pressures than in bulk. The findings from this work can be used to develop and validate thermodynamic models capable of predicting the phase behavior of CO_2 under nanoconfinement.

1. Introduction

The ever-rising level of carbon dioxide (CO_2) emission has sparked significant interest in developing new methods and materials aimed at sequestration and storage of this fluid before and after its release into the atmosphere [1,2]. Among the various techniques for CO_2 capture, adsorption has gained prominence due to its capacity to generate high-purity gas streams while consuming minimal energy and offering direct operational processes [3,4]. As a result, several adsorbent materials have been specifically engineered to fulfill this objective [5–7]. Mesoporous silicas such as MCM-41, MCM-48, and SBA-15 are considered exceptional substrates due to their consistent hexagonal pore arrangement, substantial surface area, and remarkable thermal resilience. Nevertheless, the unmodified silica surface has limited affinity towards the CO_2 molecule, primarily via physisorption interactions. Consequently, it may exhibit a reduced adsorption capability, particularly when exposed to low concentrations of CO_2 in the gas stream [8–11].

Storage of CO_2 in porous materials significantly depends on the pore space properties. Because of confinement, the thermophysical properties of fluids in nano-scale pores differ from those observed in bulk spaces. A comprehensive understanding of such behavior is essential for various industrial applications and processes, including catalysis [12–14], the pharmaceutical industry [15,16], carbon dioxide sequestration [17,18], nanomaterial analysis [19,20], pollution mitigation [21,22], as well as hydrocarbon extraction and recovery in the oil and gas sector [23,24]. To date, many well-known phenomena have been observed in confined pure fluids. For instance, empirical evidence has shown that phase transitions inside a confined space occur isothermally at lower pressures or isobarically at higher temperatures than in bulk spaces [25,26,35–39,27–34]. In the context of nanoporous media, it is essential to consider the size of the molecule and mean free path with respect to pore size when dealing with pores of diameters smaller than 100 nm and larger than 2 nm [28,30]. In this particular size range, the decrease in distances between molecules caused by confinement leads to an increase in intermolecular forces. As a result, the phase behavior is significantly influenced by the interactions between fluid molecules, as seen in bulk systems, as well as interactions between the pore walls and fluid [28,33]. While it is qualitatively understood that factors such as pore size, pore shape, interconnectivity, pore wall roughness, surface wettability, and fluid composition, along with molecular size, collectively influence the physical behavior of confined fluids, a comprehensive quantitative understanding of the relative impacts of each of these characteristics remains an ongoing challenge [28].

The scientific communities of various disciplines have shown continuous interest in the details of carbon dioxide (CO_2) adsorption on nanoporous materials, particularly micro- and mesoporous silicas. This heightened interest is driven by the urgent challenges associated with CO_2 capture and sequestration and the utilization of CO_2 in enhanced gas recovery from shales and coal formations. Carbon dioxide (CO_2) is an appropriate probe molecule for the characterization of microporous materials [40,41]. This is because CO_2 can be adsorbed at ambient temperatures, which enables quicker diffusion rates and penetration into the smallest micropores that are often inaccessible to cryogenic adsorbates. In recent years, significant advancements in adsorption measurement technology have allowed for the dependable execution of gas adsorption studies [42]. These experiments are now conducted using automated adsorption instruments, which can operate at high pressures and temperatures, specifically with carbon dioxide as the adsorbate. Consequently, the pore size distributions (PSDs) throughout the whole micro- and mesopore spectrum can be evaluated from a single experimental isotherm of CO_2 [43,44]. A comprehensive understanding of CO_2 adsorption and phase behavior in micro- and mesopores, particularly the processes of capillary condensation and hysteresis, is crucial for accurately interpreting the isotherms. The primary obstacle is the lack of knowledge of condensation pressure occurring within the domain characterized by hysteresis and its relationship with variables of pore size and temperature [6,45].

Most examples available in the literature adopted numerical techniques to probe the phase characteristics of CO_2 in confined spaces. There is a scarcity of research efforts aimed at experimentally examining the impacts of confinement. Moreover, the literature shows that most studies were conducted in the low-temperature range due to the high saturation pressures of CO_2 at high temperatures and limited to a narrow spectrum of pore sizes. This study examines the effect of confinement on the phase behavior of CO_2 in mesoporous silica (MCM-41) at varying temperatures, ranging from -23.1 to 20 °C, and pore sizes, using a novel gravimetric apparatus [25,26,37,46]. The subsequent sections of the article are structured in the following manner. Section 2 provides the specifications of the nanoporous material MCM-41 and carbon dioxide, used as adsorbent and adsorbate, respectively, in this work. In addition, it describes the methodologies for performing the experiments along with details about the apparatus, experimental procedure, and data collection and processing. The results of the experiments, including the experimentally measured adsorption and desorption isotherms for two pore sizes (6 and 10 nm), are presented and discussed in Section 3. Finally, this work is concluded in Section 4 with some final remarks.

2. Materials and Methods

2.1 Materials:

The model mesoporous material, MCM-41, consists of nanopores with precisely controlled dimensions and configurations. While these pores are conventionally characterized as circular cross-sections, several investigations have confirmed them to possess a hexagonal shape [25,26,47]. In this study, MCM-41 samples with consistent hexagonal pore shapes and varying pore diameters of 6 and 10 nm (from Glantreo, Ltd.) were employed. High purity research grade carbon dioxide (99.99% purity) was acquired from Airgas, Inc. and used as the adsorbate fluid for the experiments.

In order to characterize these samples and gather insights into their pore size, shape, and volume, a sophisticated imaging technique of Transmission Electron Microscopy (TEM) was used. On the other hand, the Brunauer-Emmett-Teller (BET) and the Barrett-Joyner-Halenda (BJH) methods were utilized to determine the specific surface area (SSA) and pore volume of each sample, as presented in **Table 1** [48,49,50]. The TEM-generated images, as depicted in **Figure 1a**, exhibited a distinctive honeycomb-like structure, confirming that the pores in all MCM-41 samples utilized in this research possessed hexagonal cross-sections, which is in contrast to the conventional assumption of circular cross-sections. Furthermore, these pores exhibited a high degree of order, uniformity, and a lack of interconnections, thereby substantiating the material's suitability for the purposes of this investigation. The TEM micrograph illistrated in **Figure 1b** shows the pore space geometry of MCM-41 material from the pore axis direction. It demonstrates the presence of distinct and evenly spaced pores oriented vertically along the channels. This data confirms the presence of discrete arrays of mesoporous channels in the MCM-41 material.

Table 1. BET and BJH analysis of the used MCM-41 samples.

Sample	**Sample weight (g)**	**Outgas Temperature (°C)**	**Outgas Time (hrs)**	**BET Surface Area (m^2/g)**	**BJH Surface Area (m^2/g)**	**BJH Pore Radius (Å)**	**BJH Pore Volume (cc/g)**
MCM-41 (6 nm)	0.1020	250	6	619.575	383.497	28.006	0.569
MCM-41 (10 nm)	0.0851	250	6	657.042	494.860	47.668	1.224

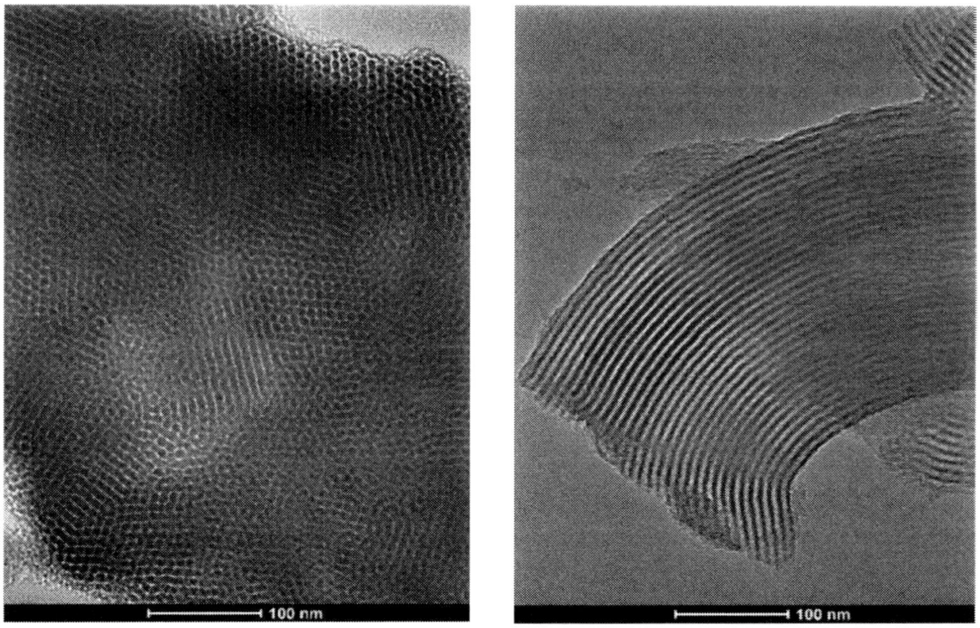

(a) (b)

Figure 1. High-resolution TEM images of the 10 nm MCM-41.

2.2 Experimental Setup and Procedure:

The adsorption and desorption behaviors were investigated using a gravimetric experimental arrangement shown in **Figure 2** [25,26,37,46]. The apparatus consisted of a Thermotron environmental chamber with highly precise temperature control and an accuracy of ±0.1 °C. Four core holders containing samples of nanoporous materials were suspended inside the chamber via thin metal cables connected to Mettler Toledo XPE505C mass comparators. The balances, with an accuracy of 0.00001 g, measured and recorded the mass of the core holders during the experiment. The fluid being examined was fully confined inside the core holder. As a result, no buoyancy force adjustments were required. Notably, this setup allows tests to be conducted for broad ranges of temperatures and pressures on significantly large quantities of porous materials, typically ranging from 6 to 8 grams. To minimize errors in mass readings originating from vibrations in the system's mechanical components, the mass comparators were positioned on an anti-vibration table supported by a sturdy frame.

For the adsorption part of the experiments, carbon dioxide was introduced directly into the core holders either from a pressurized gas cylinder or via a high-pressure Quizix-6000 series pump (Chandler Engineering). In contrast, during the desorption phase, the fluid was withdrawn from the system in small doses using a vacuum pump. The amount of injected and retracted fluid was controlled by pneumatic Vindum valves, specifically designed to open for certain durations programmed in the LabVIEW software. Hastelloy metal tubing with a 1/8" diameter was utilized to transport the fluid from the source to the environmental chamber. However, inside the chamber, flexible 1/16" peek tubing connected the core holders to the system. Rosemount pressure transducers (Emerson) and TM 101 (Leybold) vacuum gauges were employed to record the system pressure in real-time during the adsorption and desorption tests and vacuuming process, respectively. A customized data-acquisition system and a LabVIEW software program were used for efficient data collection and processing. **Figure 2** depicts a schematic representation of the experimental setup used in this study.

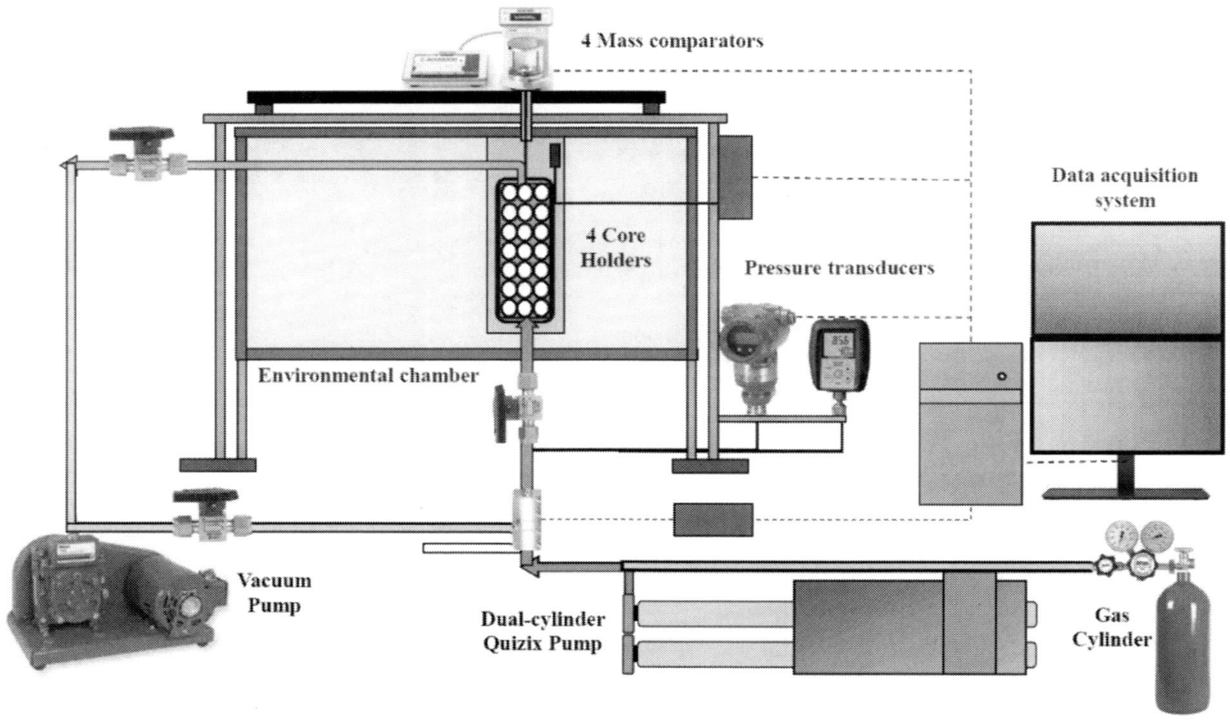

Figure 2. The schematic diagram of the experimental setup used in this study.

3. Results and Discussion

Figures 3 and **4** show experimentally measured adsorption and desorption isotherms for two distinct pore sizes, 6 and 10 nm, for a range of temperatures (-23.1, -10, 0, 10, and 20 °C). It should be noted that all tests were conducted in the subcritical regime of carbon dioxide (CO_2). The vertical dashed lines intersecting the isotherms indicate the bulk saturation pressure data from the National Institute of Standards and Technology (NIST) for each temperature. Meanwhile, the vertically dotted lines originating from the bottom axis depict the capillary condensation pressure, which is calculated by the Lorentzian fitting method. In this approach, the rate of change in mass with respect to pressure, is fitted to a Lorentzian function curve. The peak of which corresponds to the fastest rate of mass adsorption, i.e., capillary condensation. A comprehensive description of the Lorentzian technique used for the determination of capillary condensation pressure can be found in other papers published by our research group [25,26,37,46].

The inferences from the figures presented below reveal the impact of temperature on the confined phase behavior of CO_2. It is evident from the results that the phase transition due to confinement becomes more pronounced at low temperatures, leading to a more distinctive condensation of the confined phase, commonly referred to as capillary condensation. Additionally, the quantity of gas condensed within these tight spaces increases with decreasing temperature. For instance, at -23 and -10 °C, capillary condensation commences more abruptly, resulting in a larger amount of gas condensed within the confined spaces compared to the cases of higher temperatures (0, 10, 20 °C). It is crucial to acknowledge that temperature significantly affects the density profile of the fluid, and these effects are also reflected in the observed isotherms.

In addition, the results demonstrate that the degree of steepness of the adsorption curve in the capillary condensation region decreases at ascending temperature. This suggests that carbon dioxide (CO_2) reaches a supercritical state inside the nanopores at temperatures much lower than the critical point of CO_2 in bulk. Under these conditions, there is a simultaneous presence of the supercritical phase inside the nanopores and the subcritical gaseous phase in the bulk spaces (larger volumes). As expected, this behavior is prevalent when temperatures are in proximity to the bulk critical point, for example, 20 °C for carbon dioxide. At this temperature, the phenomenon of phase change is less distinct compared to the cases at lower temperatures. Furthermore, the isotherms indicate that the density of the supercritical confined phase is lower than that of the confined liquid phase at lower temperatures but higher than the density of the gas phase at the same temperature (20 °C). The obtained results conform to the findings available in the literature ([28,32,59,51–58]), asserting that the confined critical point for most gases occurs at temperatures lower than the bulk critical point. For carbon dioxide, the bulk critical point is 31 °C.

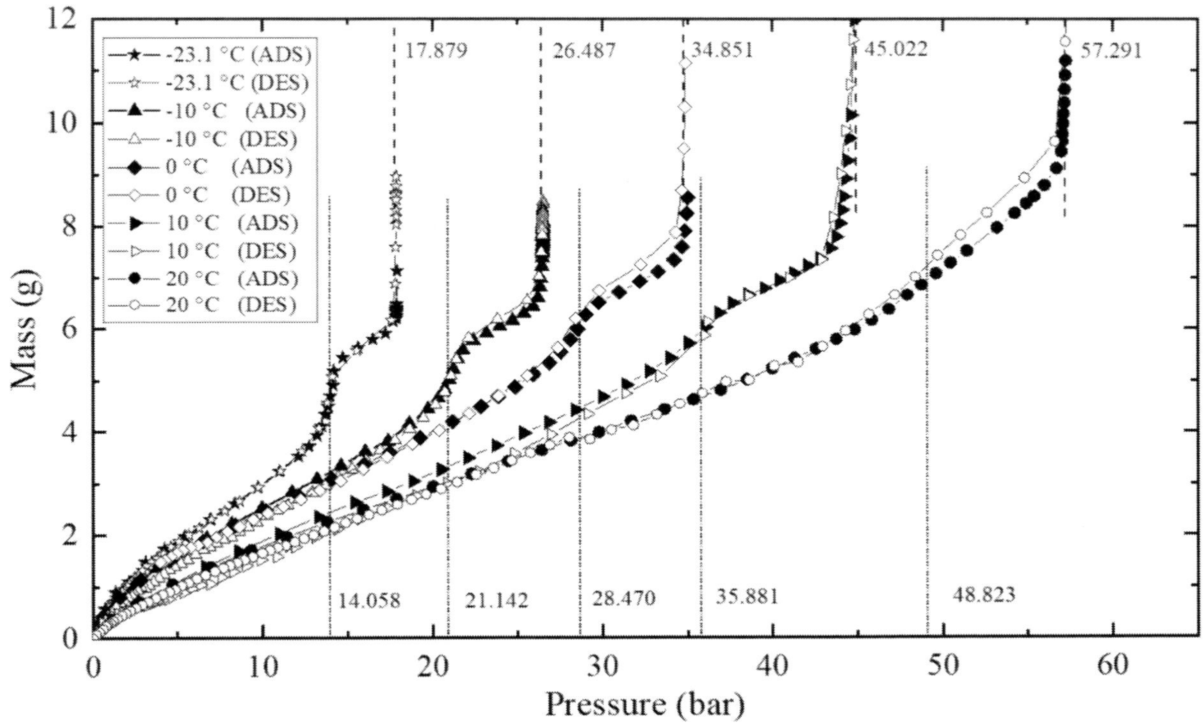

Figure 3. The experimentally measured adsorption and desorption isotherms of CO_2 in MCM-41 of 6 nm pore size at varying temperatures.

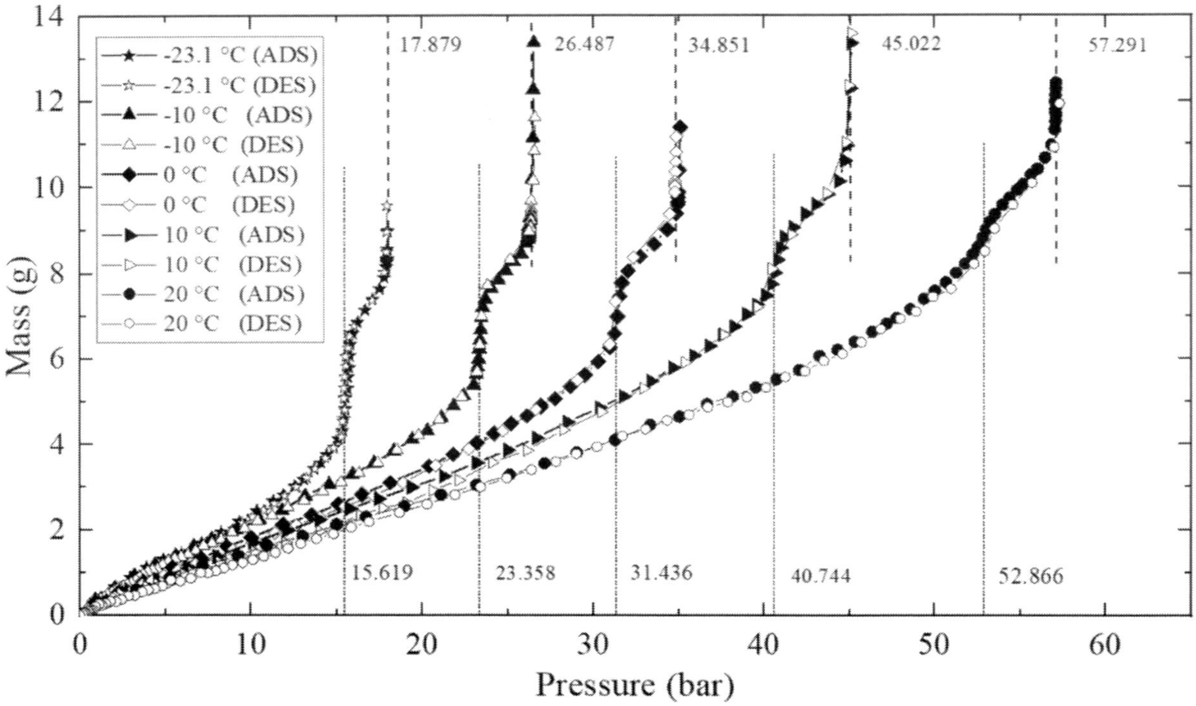

Figure 4. The experimentally measured adsorption and desorption isotherms of CO_2 in MCM-41 of 10 nm pore size at varying temperatures.

4. Conclusion

In summary, the data presented depicts the significant impact of temperature on the confined phase behavior of CO_2 in nanopores. As temperatures decrease, the manifestation of confined phase condensation, known as capillary condensation, becomes notably more pronounced. This results in a more abrupt onset of capillary condensation at lower temperatures, leading to a higher volume of gas condensed within the confined spaces. In contrast, at higher temperatures, such as 0, 10, and 20 °C, this behavior is less pronounced. Furthermore, the results indicate a decrease in the steepness of the adsorption curve within the capillary condensation region as temperature rises. This phenomenon suggests the presence of a supercritical phase within the nanopores at temperatures below the bulk critical point of CO_2. In such cases, both supercritical and subcritical gaseous phases coexist within the confined spaces and bulk spaces. This behavior is particularly evident near the bulk critical point, primarily at 20 °C for CO_2. At this temperature, the phase transition is less distinct compared to lower-temperature scenarios. Additionally, the isotherms reveal that the density of the supercritical confined phase is higher than that of the gas phase at the same temperature (20 °C) but lower than the density of the confined liquid phase observed at lower temperatures. In conclusion, the results underscore the remarkable influence of temperature on the confined phase behavior of CO_2 in nanoporous materials and contribute valuable insights to our understanding of adsorption and desorption processes at the nano-scale.

The authors express sincere appreciation for the financial support provided by Hess Corporation and the University of Wyoming throughout this research.

References

1. Song, C. Global challenges and strategies for control, conversion and utilization of CO2 for sustainable development involving energy, catalysis, adsorption and chemical processing. *Catal. Today* **115**, 2–32 (2006).

2. Yang, H. *et al.* Progress in carbon dioxide separation and capture: A review. *J. Environ. Sci.* **20**, 14–27 (2008).

3. It, I. *et al.* Carbon capture from air, is it an option? *AIDS Read.* **18**, 292 (2008).

4. Raghuvanshi, S. P., Chandra, A. & Raghav, A. K. Carbon dioxide emissions from coal based power generation in India. *Energy Convers. Manag.* **47**, 427–441 (2006).

5. Sun, H., Zhao, H., Qi, N. & Li, Y. Molecular Insights into the Enhanced Shale Gas Recovery by Carbon Dioxide in Kerogen Slit Nanopores. *J. Phys. Chem. C* **121**, 10233–10241 (2017).

6. Dantas, S., Struckhoff, K. C., Thommes, M. & Neimark, A. V. Phase Behavior and Capillary Condensation Hysteresis of Carbon Dioxide in Mesopores. *Langmuir* **35**, 11291–11298 (2019).

7. Milner, P. J. *et al.* A Diaminopropane-Appended Metal-Organic Framework Enabling Efficient CO2 Capture from Coal Flue Gas via a Mixed Adsorption Mechanism. *J. Am. Chem. Soc.* **139**, 13541–13553 (2017).

8. Bhagiyalakshmi, M., Yun, L. J., Anuradha, R. & Jang, H. T. Synthesis of chloropropylamine grafted mesoporous MCM-41, MCM-48 and SBA-15 from rice husk ash: Their application to CO2 chemisorption. *J. Porous Mater.* **17**, 475–484 (2010).

9. Thommes, M., Köhn, R. & Fröba, M. Sorption and pore condensation behavior of pure fluids in mesoporous MCM-48 silica, MCM-41 silica, SBA-15 silica and controlled-pore glass at temperatures above and below the bulk triple point. *Appl. Surf. Sci.* **196**, 239–249 (2002).

10. Liu, Y. *et al.* Performance of mesoporous silicas (MCM-41 and SBA-15) and carbon (CMK-3) in the removal of gas-phase naphthalene: Adsorption capacity, rate and regenerability. *RSC Adv.* **6**, 21193–21203 (2016).

11. Coasne, B., Galarneau, A., Di Renzo, F. & Pellenq, R. J. M. Gas adsorption in mesoporous micelle-templated silicas: MCM-41, MCM-48, and SBA-15. *Langmuir* **22**, 11097–11105 (2006).

12. Xu, S. *et al.* Surface barriers to mass transfer in nanoporous materials for catalysis and separations. *Chem. Soc. Rev.* **52**, 3991–4005 (2023).

13. Gholami, A., Pourfayaz, F. & Maleki, A. Recent Advances of Biodiesel Production Using Ionic Liquids Supported on Nanoporous Materials as Catalysts: A Review. *Front. Energy Res.* **8**, (2020).

14. Sneddon, G., Greenaway, A. & Yiu, H. H. P. The potential applications of nanoporous materials for the adsorption, separation, and catalytic conversion of carbon dioxide. *Adv. Energy Mater.* **4**, 1–19 (2014).

15. Duan, R., Xia, F. & Jiang, L. Constructing tunable nanopores and their application in drug delivery. *ACS Nano* **7**, 8344–8349 (2013).

16. Thommes, M. & Cychosz, K. A. Physical adsorption characterization of nanoporous materials:

Progress and challenges. *Adsorption* **20**, 233–250 (2014).

17. Chen, C. & Ahn, W. S. CO 2 adsorption on LTA zeolites: Effect of mesoporosity. *Appl. Surf. Sci.* **311**, 107–109 (2014).

18. Belmabkhout, Y., Serna-Guerrero, R. & Sayari, A. Adsorption of CO2 from dry gases on MCM-41 silica at ambient temperature and high pressure. 1: Pure CO2 adsorption. *Chem. Eng. Sci.* **64**, 3721–3728 (2009).

19. Datar, A., Chung, Y. G. & Lin, L. C. Beyond the BET Analysis: The Surface Area Prediction of Nanoporous Materials Using a Machine Learning Method. *J. Phys. Chem. Lett.* **11**, 5412–5417 (2020).

20. Keshavarz, L., Ghaani, M. R., Saremi, O. & English, N. J. Characterization of Nanoporous Materials. *Eng. Mater.* 319–351 (2022) doi:10.1007/978-3-030-85397-6_10.

21. Vidal-Limon, A., García Suárez, P. C., Arellano-García, E., Contreras, O. E. & Aguila, S. A. Enhanced Degradation of Pesticide Dichlorophen by Laccase Immobilized on Nanoporous Materials: A Cytotoxic and Molecular Simulation Investigation. *Bioconjug. Chem.* **29**, 1073–1080 (2018).

22. Thomas, K. M. Adsorption and desorption of hydrogen on metal-organic framework materials for storage applications: Comparison with other nanoporous materials. *Dalt. Trans.* 1487–1505 (2009) doi:10.1039/b815583f.

23. Gimenez, R., Berli, C. L. A. & Bellino, M. G. Oil Recovery from Nanoporous Media via Water Condensation. *Adv. Mater. Interfaces* **6**, 1–8 (2019).

24. Fan, J. C. *et al.* Molecular mechanism of viscoelastic polymer enhanced oil recovery in nanopores. *R. Soc. Open Sci.* **5**, (2018).

25. Sharma, K. V., Alloush, R. M. & Piri, M. Confined phase behavior of ethane in nanoporous media: An experimental investigation probing the effects of pore size and temperature. *Microporous Mesoporous Mater.* **351**, 112459 (2023).

26. Alloush, R. M., Sharma, K. V. & Piri, M. Capillary condensation and evaporation of n-butane and iso -butane in nanoporous media : Experimental study of the effects of pore size and temperature using an automated gravimetric apparatus. *J. Mol. Liq.* **364**, 119894 (2022).

27. Gelb, L. D., Gubbins, K. E., Radhakrishnan, R. & Sliwinska-Bartkowiak, M. Phase separation in confined systems. *Reports Prog. Phys.* **62**, 1573–1659 (1999).

28. Barsotti, E., Tan, S. P., Saraji, S., Piri, M. & Chen, J. H. A review on capillary condensation in nanoporous media: Implications for hydrocarbon recovery from tight reservoirs. *Fuel* **184**, 344–361 (2016).

29. Barsotti, E., Qin, Z. & Piri, M. In Situ Investigation of Fluid-Rock Interactions at Ångstrom Resolution. *J. Geophys. Res. Solid Earth* **126**, (2021).

30. Barsotti, E., Lowry, E., Piri, M. & Chen, J. H. Using Capillary Condensation and Evaporation Isotherms to Investigate Confined Fluid Phase Behavior in Shales. in *E3S Web of Conferences* vol. 146 (EDP Sciences, 2020).

31. Barsotti, E., Tan, S. P., Piri, M. & Chen, J. H. Phenomenological Study of Confined Criticality:

Insights from the Capillary Condensation of Propane, n -Butane, and n -Pentane in Nanopores. *Langmuir* **34**, 4473–4483 (2018).

32. Tan, S. P., Barsotti, E. & Piri, M. Criticality of Confined Fluids Based on the Tensile Strength of Liquids. *Ind. Eng. Chem. Res.* **59**, 10673–10688 (2020).

33. Tan, S. P., Barsotti, E. & Piri, M. Application of material balance for the phase transition of fluid mixtures confined in nanopores. *Fluid Phase Equilib.* **496**, 31–41 (2019).

34. Luo, S., Lutkenhaus, J. L. & Nasrabadi, H. *SPE-175095-MS Experimental Study of Confinement Effect on Hydrocarbon Phase Behavior in Nano-Scale Porous Media Using Differential Scanning Calorimetry*. http://onepetro.org/speatce/proceedings-pdf/15atce/3-15atce/d031s043r003/1381706/spe-175095-ms.pdf/1 (2015).

35. Sedghi, M. & Piri, M. Capillary condensation and capillary pressure of methane in carbon nanopores: Molecular Dynamics simulations of nanoconfinement effects. *Fluid Phase Equilib.* **459**, 196–207 (2018).

36. Alam, M. A., Clarke, A. P. & Duffy, J. A. Capillary condensation and desorption of binary mixtures of N2-Ar confined in a mesoporous medium. *Langmuir* **16**, 7551–7553 (2000).

37. Barsotti, E., Saraji, S., Tan, S. P. & Piri, M. Capillary Condensation of Binary and Ternary Mixtures of n-Pentane-Isopentane-CO2 in Nanopores: An Experimental Study on the Effects of Composition and Equilibrium. *Langmuir* **34**, 1967–1980 (2018).

38. Vardhan Sharma, K., Alloush, R. M., Al-Garadi, K. & Piri, M. The effect of nanoconfinement on the phase behavior of ethane/n-propane binary mixture: An experimental study at varying pore sizes and compositions. *35th Int. Symp. Soc. Core Anal.* 1–12 (2022).

39. Lowry, E. & Piri, M. Effects of chemical and physical heterogeneity on confined phase behavior in nanopores. *Microporous Mesoporous Mater.* **263**, 53–61 (2018).

40. Garrido, J. & Torregrosa, R. Use of N_2 vs .CO_2 in the Characterization of Activated Carbons. *Langmuir* **3**, 76–81 (1987).

41. Rodríguez-Reinoso, F., Garrido, J., Martín-Martínez, J. M., Molina-Sabio, M. & Torregrosa, R. The combined use of different approaches in the characterization of microporous carbons. *Carbon N. Y.* **27**, 23–32 (1989).

42. Silvestre-Albero, J., Silvestre-Albero, A., Rodríguez-Reinoso, F. & Thommes, M. Physical characterization of activated carbons with narrow microporosity by nitrogen (77.4 K), carbon dioxide (273 K) and argon (87.3 K) adsorption in combination with immersion calorimetry. *Carbon N. Y.* **50**, 3128–3133 (2012).

43. Nguyen, H. G. T. *et al.* Correction to: A reference high-pressure CO2 adsorption isotherm for ammonium ZSM-5 zeolite: results of an interlaboratory study (Adsorption, (2018), 24, 6, (531-539), 10.1007/s10450-018-9958-x). *Adsorption* **24**, 703 (2018).

44. Nguyen, H. G. T. *et al.* A reference high-pressure CO2 adsorption isotherm for ammonium ZSM-5 zeolite: results of an interlaboratory study. *Adsorption* **24**, 531–539 (2018).

45. Rios, R. V. R. A., Silvestre-Albero, J., Sepúlveda-Escribano, A., Molina-Sabio, M. & Rodríguez-Reinoso, F. Kinetic restrictions in the characterization of narrow microporosity in carbon materials. *J. Phys. Chem. C* **111**, 3803–3805 (2007).

46. Piri. *Hydrocarbons Adsorption on Templated Mesoporous Materials : Effect of the Pore Size , Geometry and Surface Chemis try. Materials Research Society Symposium Proceedings* vol. 29 (2013).

47. Kruk, M. *et al.* Determination of Pore Size and Pore Wall Structure of MCM-41 by Using Nitrogen Adsorption, Transmission Electron Microscopy, and X-ray Diffraction. *J. Phys. Chem. B* **104**, 292–301 (2000).

48. Brunauer, S., Emmett, P. H. & Teller, E. Adsorption of gases in multimolecular layers. *J. Am.Chem.Soc.* **60**, 309–319 (1938).

49. Barrett, P., Joyner, G. & Halenda, P. P. The Determination of Pore Volume and Area Distributions in Porous Substances. I. Computations from Nitrogen Isotherms. (1951).

50. Joyner, G., Skold, R. & Barrett, P. The determination of pore volume and area distributions in porous substances. II. Comparison between nitrogen isotherm and mercury porosimeter methods. **3**, (1951).

51. Thommes, M. & Findenegg, G. H. Pore Condensation and Critical-Point Shift of a Fluid in Controlled-Pore Glass. *Langmuir* **10**, 4270–4277 (1994).

52. Melnichenko, Y. B., Wignall, G. D., cole d, Frielinghaus, H. & Bulavin L. Liquid-gas critical phenomena under confinement: Small-angle neutron scattering studies of CO2 in aerogel. *J. Mol. Liq.* (2005).

53. Wong, A. P. Y. & Chan, M. H. W. *Liquid-Vapor Critical Point of He in Aerogel*. vol. 65 (1990).

54. Tan, S. P., Qiu, X., Dejam, M. & Adidharma, H. Critical Point of Fluid Confined in Nanopores: Experimental Detection and Measurement. *J. Phys. Chem. C* **123**, 9824–9830 (2019).

55. Wong, ' ', A. P. Y., Kim, S. B., Goldburg, W. I. & Chan ', M. H. W. *Phase Separation, Density Fluctuation, and Critical Dynamics of Nz in Aerogel*. vol. 70 (1993).

56. Morishige, K. & Shikimi, M. Adsorption hysteresis and pore critical temperature in a single cylindrical pore. *J. Chem. Phys.* **108**, 7821–7824 (1998).

57. Qiu, X., Tan, S. P., Dejam, M. & Adidharma, H. Experimental Study on the Criticality of a Methane/Ethane Mixture Confined in Nanoporous Media. *Langmuir* **35**, 11635–11642 (2019).

58. Wee Tan, H. & Beamish, J. *Liquid}vapor critical behavior of neon con"ned in aerogel.* (2000).

59. Salahshoor, S., Fahes, M. & Teodoriu, C. A review on the effect of confinement on phase behavior in tight formations. *J. Nat. Gas Sci. Eng.* **51**, 89–103 (2018).

84e Revitalize Conventional Legacy OIL Fields in North Dakota with Waterflooding and CO_2 EOR

Tao Jiang, Lu Jin, Stephen Guillot, Michael Warmack, James A. Sorensen, John A. Hamling, Xincheng Wan

There are 280–630 MMbbl of incremental oil recovery potential through the use of tertiary CO_2 enhanced oil recovery (EOR) in 86 North Dakota unitized conventional oil fields (KLJ, 2014). A joint industry research program led by the Energy & Environmental Research Center (EERC) was established to investigate the effectiveness of implementing waterflooding and optimizing it as a precursor for CO_2 EOR and to develop cost-effective pathways for a legacy oil field, Foreman Butte Field, as a prototype for revitalizing the analogous conventional oil fields in North Dakota to prolong the operational lifetime of the fields with low carbon intensity.

A horizontal well was first drilled into the Flat Lake subinterval of the Foreman Butte Field in 2003, and wireline logs suggested a 12-foot section of limestone within the Flat Lake subinterval contained an average of 5% porosity with 70% hydrocarbon saturation (Nordeng, 2007). Production data from 40 subsequently drilled horizontal wells on 640-acre spacing exhibited rapid production declines and production of more water than oil without waterflooding. A legacy field like this typically lacks characterization data for detailed geologic modeling, so additional strategic characterization data were acquired to improve understanding of the reservoir and optimize production. The presence of the natural fracture networks from the newly collected cores confirmed in assessing this type of reservoir. Thus a high-efficiency fracture modeling approach, embedded discrete fracture modeling (EDFM), was used in numerical simulation to capture the complex fracture networks and heterogeneity of the reservoir. This approach helped to improve the accuracy of history matching and predict the performance of waterflooding and CO_2 EOR with higher confidence.

Implementing a waterflood after primary production usually presents the most cost-effective approach to increasing oil recovery in terms of additional oil produced per dollar of capital investment. The development of a CO_2 EOR project after waterflooding could further recover a significant volume of oil. Both water and CO_2 flooding experiments were conducted in this study, with results suggesting that a considerably higher oil recovery could be achieved by CO_2 EOR rather than waterflooding in this field. However, it would require much higher capital investment for surface facilities and CO_2 sourcing and transportation in the field application process.

North Dakota has permitted a few carbon capture and storage (CCS) projects where the captured CO_2 could be readily available for EOR purposes, but building pipeline systems accessible to those sources may also require obtaining new separate right-of-way agreements to the legacy field. Though the CO_2 sourcing, transportation of the sourced CO_2, and the expanded separation facilities for recycling the produced CO_2 pose a higher cost and could be offset by certain regulatory incentives, such as federal- and state-level tax credits and tax exemptions, a necessary economic evaluation with detailed geologic modeling and reservoir simulation studies can provide a basis for such project implementation.

This research program will generate new information to enable operators, investors, regulators, and other stakeholders to make informed decisions regarding the revitalization of North Dakota's vast conventional oil resources using CO_2 capture and utilization techniques.

84f Development of Machine Learning [ML] Based Model for Predicting CO_2 Hydrate Formation Kinetics in Porous Media

Vikas Dhamu, M Fahed Qureshi, Praveen Linga

*Correspondence: praveen.linga@nus.edu.sg (P. Linga)

Carbon capture, utilization, and storage (CCUS) is the process of capturing CO_2 emissions and either using them to create products like construction materials (utilization) or permanently storing them below a suitable medium, such as depleted oil and gas reservoirs or deep-saline aquifers. Capturing CO_2 emissions from industry and injecting them into deep oceanic sediments to be stored as gas hydrates is a viable alternative CCUS technique. Gas hydrates are crystalline ice-liked compounds formed under high pressure and low-temperature conditions, already prevalent at depths of 300 m and above in the ocean sediments. In this work, experimentally, the formation kinetics of CO_2 hydrates in different sediments have been investigated. An ML-based kinetic model has been trained to predict CO_2 hydrate formation kinetics in sediments using experimental data. Four parameters based on a new ML-based algorithm were proposed to predict the stochastic kinetics nature of CO_2 hydrate formed in the presence of sediments. Out of four parameters, two were taken from the published literature studies.

The experiments were carried out inside a specialized high-pressure reactor in which CO_2 was injected via injection tube directly into the sediments such as silica sand (0.5-1.5 mm) and granules sand (1.5-3mm) at 3.5 MPa and 1-2 ^oC. Then using the experimental kinetic data, a new ML-based CO_2 hydrate kinetic model was developed to predict the CO_2 hydrate formation kinetics in the sediments. The water-to-hydrate conversion was estimated to be more in the presence of mixed sand [>85 %] followed by silica sand [>70%] and granules sand [>65%]. The proposed ML-based model was trained using a total of 32843 experimental kinetics data and could predict the water-to-hydrate conversion with accuracy in terms of %AARD (<10%) for all kinds of experiments done in this study. The experimental results and proposed ML-based kinetic model will help bridge the critical knowledge gap and serve as a step forward for further developing a sustainable hydrate-based CCUS technology for the energy sector.

A Model Selection Workflow for Assimilating Time-Lapse Seismic Data in Models for Point Bar Geologic System.

Sanjay Srinivasan, Ismael Dawuda

Point bar geology is frequently observed in many target reservoirs that are used for large-scale injection and storage of CO_2. To reliably assess the containment of the injected CO_2 within the reservoir, model calibration algorithms have been developed to reduce the uncertainty associated with characterizing the geologic heterogeneity in point bar reservoirs. However, due to the complex geologic heterogeneity exhibited by point bar reservoirs, there can be significant amount of residual uncertainty even after model calibration. This can affect the reliability in predicting the reservoir storage capacity and future migration of the CO_2 plume. In this study, a model selection workflow is proposed to refine an initial suite of calibrated point bar models to create a posterior set of models. These posterior models reflect time-lapse seismic information based on observed dynamic response. The model selection workflow proceeds by: computing discrete Fréchet distances to quantify the similarity in time-lapse seismic responses obtained from a large prior ensemble of models, combining multidimensional scaling with k-means clustering to partition the models into subgroups based on their seismic responses, Performing Bayesian computations in the reduced model space to select the subgroup of models that yield response closest to the observed seismic information, and iteratively sampling the posterior models to further refine the selection of the model clusters. The final ensemble of the refined selected models can be used to assess the uncertainty in predicting the future displacement of CO_2 plume. The applicability of the workflow to a real field scenario is demonstrated using the CO_2 injection dataset for the Cranfield in Mississippi. The proposed model selection workflow has the potential to efficiently map the CO_2 plume migration and assess the uncertainty in CO_2 migration paths.

84i Expanding the Adsorbate Binding Energy Correlation from Pt Surface Site Stabilities to the Surface Site Stabilities of Other Metals

Shyama Charan Mandal, Frank Abild-Pedersen

The dehydrogenation of hydrocarbons is a complicated reaction with many reaction steps. Therefore, establishing correlations among different metals can help accelerate the discovery of efficient and cost-effective catalysts. Our α-scheme, coordination-based model can effectively and precisely predict metal binding energies (ΔE_M) with high efficiency. Additionally, we have categorized different hydrocarbon-based adsorbates according to their binding energies and surface site stability on various Pt surfaces. Furthermore, all the classes are linearly associated with adsorbate CH and it can be used as a descriptor. In this study, we have examined a range of metals, including Ag, Au, Co, Cu, Ir, Ni, Os, Pd, Pt, Rh, and Ru, and investigated to understand the correlation among them. The results of the verification of the coordination-based model demonstrate that can be obtained with high accuracy and efficiency (Figure a). Furthermore, we have identified a linear scaling relationship between the α-scheme predicted ΔE_M and the reaction descriptor, (ΔE_{CH}) (Figure b). This demonstrates that the ΔE_M exhibit a linear scaling relationship with ΔE_{CH}, for all metals. Interestingly, our findings indicate that certain metals exhibit correlations with one another under specific surface site stabilities, indicating that different metals are interrelated in a particular coordination environment. The binding energies of CH can be obtained for other metals from any metals considered in this study at a specific coordination environment. All of these are correlated with the coordination-based α-scheme model. Therefore, with this understanding of correlations, it becomes much simpler to comprehend the intricacies of hydrocarbon chemistry on various metals, and reaction energies can be obtained easily through these correlations, without the need for extensive computational calculations. These approaches can be expanded further to investigate other complex reactions. We are confident that the established correlations will prove useful for accurately and efficiently screening a vast number of catalysts.

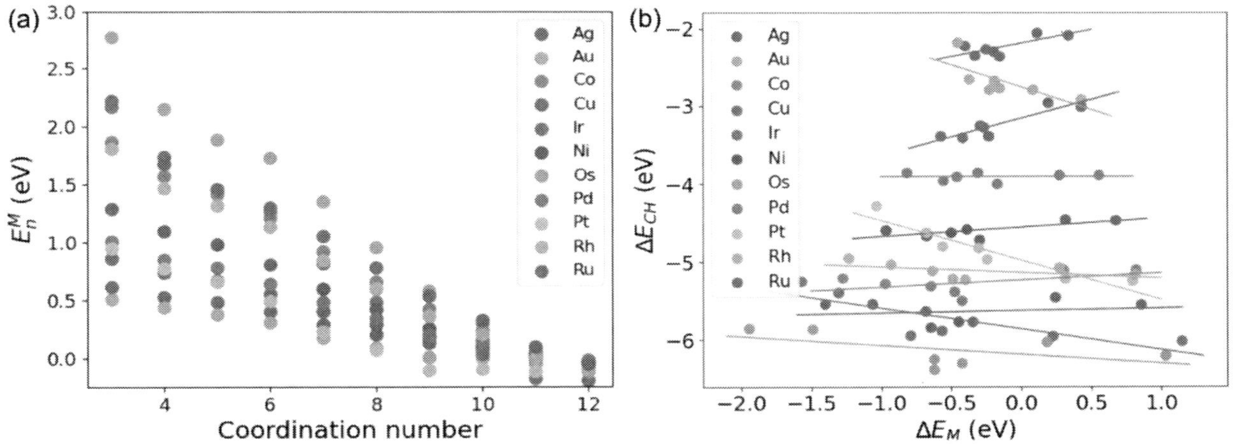

84j Predicting Surface Coverage Effects in Heterogeneous Catalysis Via an Interaction-Counting Approach

Deep Patel, Luke Roling

Accurately calculating adsorbate-adsorbate interactions is crucial for predicting in-situ properties of catalysts and reaction mechanisms at the catalyst interface. Numerous efforts have been made to accurately predict adsorbate-adsorbate interactions at the surface of transition metals, such as through the use of cluster expansion (CE) methods,[1,2] thereby improving the ability to describe reaction energetics under varying chemical potentials. Although CE has relatively high prediction accuracy (0.01 – 0.001 eV), its computational cost (often thousands of DFT calculations) remains a key barrier to its broadest implementation in high-throughput screening studies.

In this work, we provide a novel, computationally efficient (10-100 DFT calculations), and generalizable interaction-counting approach to more effectively include adsorbate interactions in catalyst screening methods. Our model enables prediction of interaction energies of simple adsorbates, including CO, O, CH_3, and CCH_3, on various transition metal surfaces with sufficient accuracies (~0.01-0.10 eV) for implementation in applications relying on general interaction correlations (e.g., microkinetic modeling). The predicted interaction parameters can be correlated to the nature of the host transition metal. Finally, we leverage the use of leave-one-out cross validation to enable our methodology to predict generalizable and reduced training sets (~15-50 configurations) for extensions to new/untrained materials. This methodology is further extended for predicting interaction energies of adsorbates with different identities, including O-CO interactions on Au(100) and Au(111) surfaces.

References:

1. 1. Sanchez, J. M. *et al. Phys. A: Stat. Mech. Appl.* **128**, 334-350 (1984).
2. 2. Schmidt, D. J. *et al. J. Chem. Theory Comput.* **8**, 264–273 (2011).
3. 3. Roling, L. T. *et al. J. Phys. Chem. C* **121**, 23002-23010 (2017).

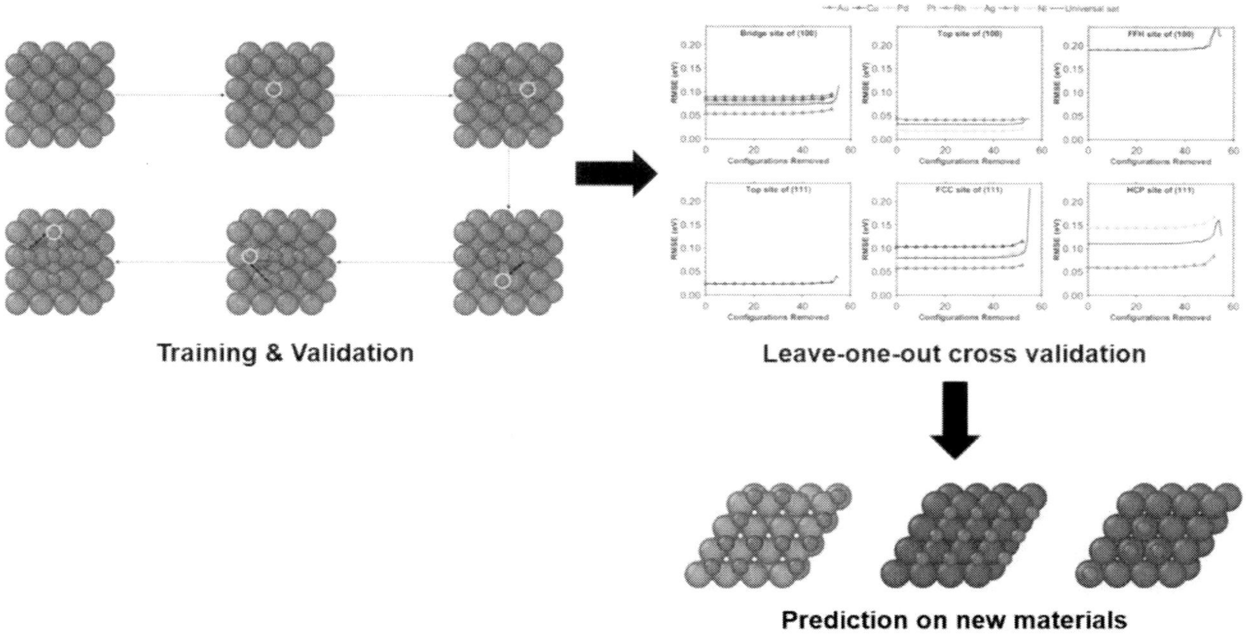
Training & Validation
Au Cu Pd Pt Rh Ag Ir Ni Universal set
Bridge site of (100)
Top site of (100)
FFH site of (100)
Top site of (111)
FCC site of (111)
HCP site of (111)
RMSE (eV)
Configurations Removed
Leave-one-out cross validation
Prediction on new materials

84k Quantification of Self-Interaction Errors in Selective Catalytic Reduction of NOx in Zeolites

Priyanka Bholanath Shukla, Selim Romero, Rajendra Zope, Tunna Baruah, Koblar Jackson, Karl Johnson

Zeolites are porous materials that are utilized as selective catalysts in various industrial reactions. To understand the adsorption energies and barrier heights of these reactions, researchers commonly use periodic density functional theory (DFT). However, DFT suffers from self-interaction errors (SIE), resulting in errors in calculated energies. For example, over-stabilization of delocalized densities in transition state structures results in underestimation of reaction barriers by about 40 kJ/mol and adsorption energies by MAE of 20 kJ/mol. This raises a need to remove SIE to make accurate property predictions. To our knowledge, the extent of SIE on reaction barriers for zeolite-catalyzed reactions has never been assessed.

The Perdew–Zunger self-interaction correction (PZSIC) and locally scaled self-interaction correction (LSIC) improve the prediction of barrier heights of chemical reactions, with LSIC giving better accuracy than PZSIC on average. These methods employ an orbital-by-orbital correction scheme to remove the one-electron SIE. Fermi–Löwdin Self-Interaction correction (FLOSIC) is a method to implement the Perdew-Zunger self-interaction corrections (PZSIC) energy using a Fermi-orbital approach.

In this study, we assess the extent of SIE in the catalytic reduction of NOx using a Cu-SSZ-13 zeolite cluster model. We compare the accuracy of commonly used DFT theory and our PZSIC and LSIC energies to study the thermodynamics and kinetics of NOx reactions. Our hypothesis is that PZSIC and LSIC can improve the predictions of reaction barriers and adsorption energies for these reactions. This study is particularly interesting because of the transition-metal based zeolites, where the transition metal changes oxidation states during the reaction, which has not been previously studied using the FLOSIC method. We have evaluated the density-driven errors by comparing the LDA and PBE energies at self-consistent PZSIC energy and functional-driven errors for these reactions. Our findings will be shared for barrier heights and reaction energies for the NOx reactions.

84m An efficient and universal solar interfacial photothermal reactor toward liquid phase oxidation

Chen Wu

Photothermal catalysis has attracted great attention owing to high solar energy utilization efficiency, and great progress has been achieved in gas-phase catalysis. However, photothermal catalysis in liquid-phase shows a limited performance, due to high specific heat capacity of liquid. Herein, an interfacial photothermal catalytic (IPC) strategy applying an IPC leaf was designed for liquid-phase catalysis using benzyl alcohol oxidation as the model reaction. On even ground, the benzaldehyde generation rate (8.7 $mmol \cdot g_{cat}^{-1} \cdot h^{-1}$) of the IPC system with Ce-doped MnO_2 as photothermal catalyst was almost 4.8 times and 3 times of bulk photothermal catalysis and the relative bulk thermal catalysis. Experimental results proved that this excellent catalytic performance could be attributed to both the high temperature achieved and the enhanced lattice oxygen activity by solar irradiation. This IPC system also exhibited good universality and long-term durability. Our work innovatively created a green strategy to obtain fine chemicals with only solar irradiation.

Figure 1. Schematic illustration of the interfacial photothermal catalysis inspired by photosynthesis toward liquid-phase benzyl alcohol oxidation.

84n Optimization of Trickle Bed Reactor Distribution: A Computational Fluid Dynamic Approach

Madison Holly, Carly Fox, Sebastian Uribe, Muthanna Al-Dahhan

Trickle bed reactors are widely applied across different chemical and biochemical industries. Despite the vast applications and uses of these systems, there is still a need to optimize these reactors in order to enhance their throughput and to enable the development of new processes. The present study is conducted to advance the knowledge in the optimization of the phases distribution in a TBR. The first place the study looks into is the distribution plate which is responsible for premixing the gas and liquid going into the reactor. Ensuring the distribution plate produces a homogeneous mixture allows for an enhanced throughput of the reactor. In this study, the distributor plate design over the phase's distribution was analyzed using Computational Fluid Dynamic techniques. STAR CCM+ was utilized to run simulations of the distributor plate to determine the maldistribution coefficient within the trickle bed reactor; a volume of fluids approach was used to approximate the fluid interactions. These simulations were validated by comparing the predictions made by the simulation against the experimental data that was measured previously.

84o The Design, Preparation of Metal-Organic Materials and Their Applications in Green Synthesis of API

Yajing Shen

Catalysts play the important role in the precise preparation of important Active Pharmaceutical Ingredient (API) with high efficiency and low waste pollution. Current catalytic system often have drawbacks like low activity, poor selectivity, and poor stability and reusability. Considering the high activity, selectivity, modifiability and adjustability of well-defined homogeneous catalysts as well as taking advantage of the large surface area, high porosity and low density of porous materials, a series of porous organometallic materials have been designed and prepared via hyper-crossing and ligand-assisted pyrolysis strategies. The morphology, structure, composition and properties of these porous materials were characterized by various analytical methods. The performance of these solid catalysts were investigated in hydrogenation, oxidation, reductive amination and formylation reactions for API synthesis, and the structure-activity relationship, reaction mechanism were deeply studied via in-situ FT-IR, NMR, XPS, XAS and so on. The catalytic materials exhibited superior activity than homogeneous precuesors and the record turnover number (TON) of 1581588 could be achieved in the formylation reaction with CO2. A series of control experiments as well as in-situ characterizations had demonstrated the high efficiency of these materials originated from the isolation effect, confinement effect, which could prevented the homogeneous precursors from agglomeration. Furthermore, the solid catalysts could be reused up to dozens of times, highlighting their practical potential in pharmaceutical industry.

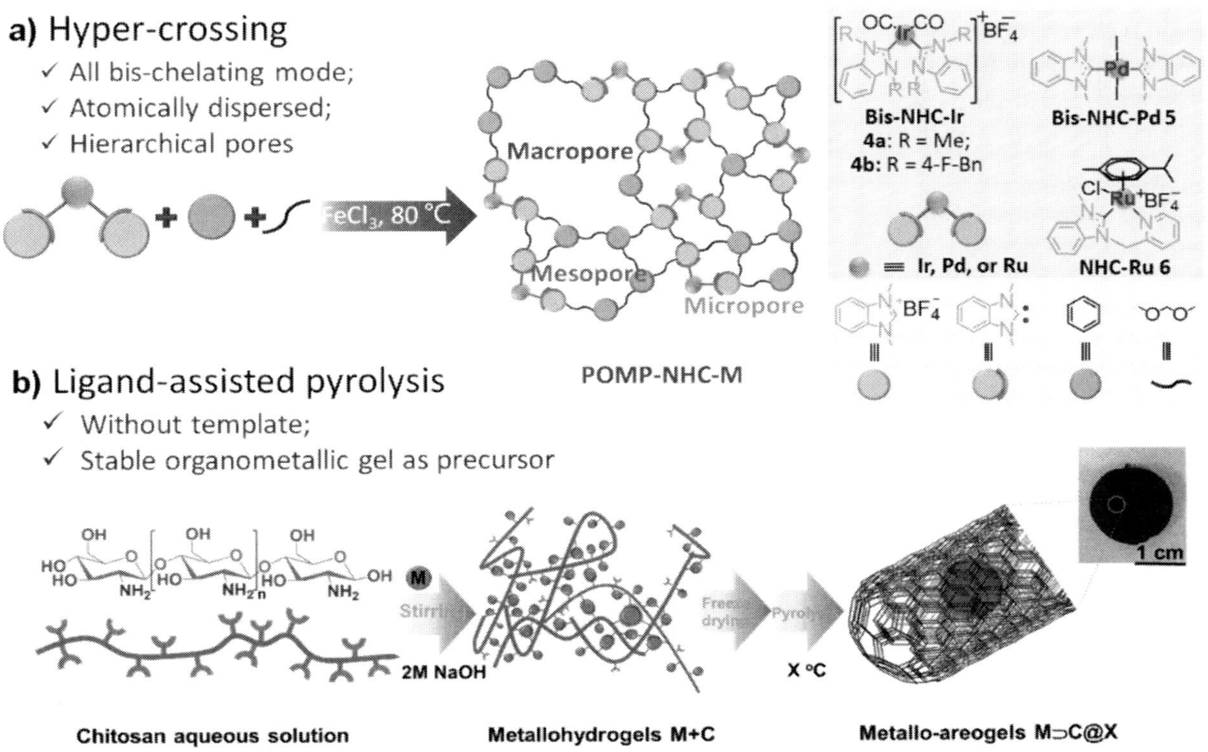

84p Engineering Bio-MOF-1 Derived Single-Atom Catalyst with a Hierarchical Porous Nanostructure for Highly Selective CO_2 Electroreduction

Yechan Lee, Byoung Joon Park, Jeong Woo Han

The increasing concentration of CO_2 in the atmosphere from consumption of large amounts of fossil fuels has caused severe environmental issues and driven the growth of sustainable and clean energy technologies. Electrochemical conversion of CO_2 into valuable fuels and chemicals under ambient conditions has been considered as a promising approach to alleviate excessive CO_2 emissions and to achieve carbon neutrality. In this regard, a promising CO_2 reduction reaction (CO_2RR) electrocatalyst is expected to exhibit high efficiency and selectivity to convert CO_2 into a desired product. Single-atom catalysts (SACs) have been recently explored as outstanding electrocatalysts for CO_2RR because of their preferred benefits, including maximum atom efficiency, unique electronic structure, and unfavorable hydrogen binding. Especially, metal-organic frameworks (MOFs) have been generally used as ideal precursors for preparing SACs, due to their simple preparation, high porosity, structural diversity, and high conductivity. Up to now, several types of MOFs, such as MOF-5, MOF-74, ZIF-8 have been utilized in the field of electrochemical CO_2RR application.

Herein, we employed bio-MOF-1 as attractive precursor for synthesizing SACs. Bio-MOF-1 is composed of Zn and adenine interconnected with biphenyldicarboxylate (BPDC) linkers. With this precursor, we finally designed and synthesized a series of nanosized bio-MOF-1 derived metal-nitrogen carbon catalysts (n-M-BMF-N-C, M = Ni, Fe, Co, Mn, Cu). These five M-N-C catalysts were then experimentally characterized and compared. Notably, electrochemical tests showed that n-Ni-BMF-N-C displayed remarkable CO Faradaic efficiency (FE) of 99 % at -1.0 V, and particularly CO FE was maintained over 90 % in a wide applied potential window, making it one of the most selective M-N-C catalysts reported. The CO_2RR selectivity trend of the n-M-BMF-N-C catalysts can be listed as following order of Cu < Mn < Co < Fe < Ni. The experimental observations have been correlated with density functional theory (DFT) simulations.

84q Strategies to Control the Microenvironment in Electrochemical CO_2 Reduction

Yaqi Cheng, Chaolong Wei, Fei Yu, Panpan Zhang, Wang Xin, Andrew Wong

The electrochemical CO_2 reduction reaction (CO_2RR) represents a critical approach toward the carbon-neutral generation of hydrocarbons. However, despite intense research efforts, the need remains to increase selectivity and activity for electrocatalysts for CO_2RR. Thus, there has been increasing interest in optimizing the microenvironment to boost activity and selectivity for CO_2 conversion.

This presentation discusses several strategies that are being pursued in my research group to optimize the microenvironment for CO_2 conversion. These strategies include: (1) control of diffusion via engineering electrode structure at micro- and nano- length scales, (2) control of diffusion via control of the localization of tandem catalysts, (3) implementation of bioelectrochemical tandem systems for CO_2 conversion, (4) implementation of MOF coatings to define the microenvironment, and (5) implementation of high-temperature non-aqueous electrolytes. Thus, this presentation surveys a wide range of ongoing research strategies to optimize the microenvironment towards more effective CO_2RR.

84r Effect of Synthesis Method on Performance of Hybrid Catalyst for Direct DME Synthesis: Flame Synthesis and Co-Precipitation.

Onochie Okonkwo, Komal Tripathi, Sonal Asthana, Yiming Xi, Sujit Modi, Kamal Pant, Pratim Biswas

The direct conversion of CO_2 to dimethyl ether (DME) is regarded as an innovative approach for upcycling CO_2. The direct DME synthesis process is catalyzed over a hybrid catalyst. Cu/ZnO/MgO and Cu/ZnO/ZrO_2 synthesized by co-precipitation and admixed with γ-Al_2O_3 and H-ZSM-5 respectively have been identified as active and selective hybrid catalysts for direct DME synthesis [1,2]. The co-precipitation process involves several steps and requires long times for catalyst synthesis. To overcome these limitations of the co-precipitation process, flame synthesis, which is a continuous single step route, is used to obtain the same catalysts. Flame synthesis provides strict control of size, surface area, morphology, and crystal phase of nanoparticles [3]. Also, flame synthesis can result in the production of metastable phases which are not accessible to conventional wet synthesis methods.

In this work, Cu/ZnO/MgO and Cu/ZnO/ZrO_2 catalyst is synthesized via flame synthesis and admixed with γ-Al_2O_3, to design the hybrid catalyst. An alternate hybrid catalyst with Cu/ZnO/MgO and Cu/ZnO/ZrO_2 synthesized via co-precipitation and admixed with γ-Al_2O_3 is also designed. The hybrid catalysts activity, selectivity, and stability for direct DME synthesis from a CO/CO_2/H_2 mixture is evaluated and compared to observe the effect of catalyst synthesis route on the performance of the hybrid catalyst. Our finding show that the hybrid catalyst manufactured via flame synthesis provides a higher activity and higher selectivity for DME over that synthesized via co-precipitation. The characterization results reveal the structural characteristics which provides insight to describe the superior performance of the flame synthesized catalyst. Altogether, flame synthesis is used to elucidate the governing aspects for high performance catalysts for direct DME synthesis.

[1] Asthana, et al, Energy & Fuels 36;5 (2022) 2673-2687.

[2] Singh, et al, Fuel 318 (2022) 123641.

[3] Okonkwo et al, Journal of the American Ceramic Society 105;2 (2022) 1481-1490.

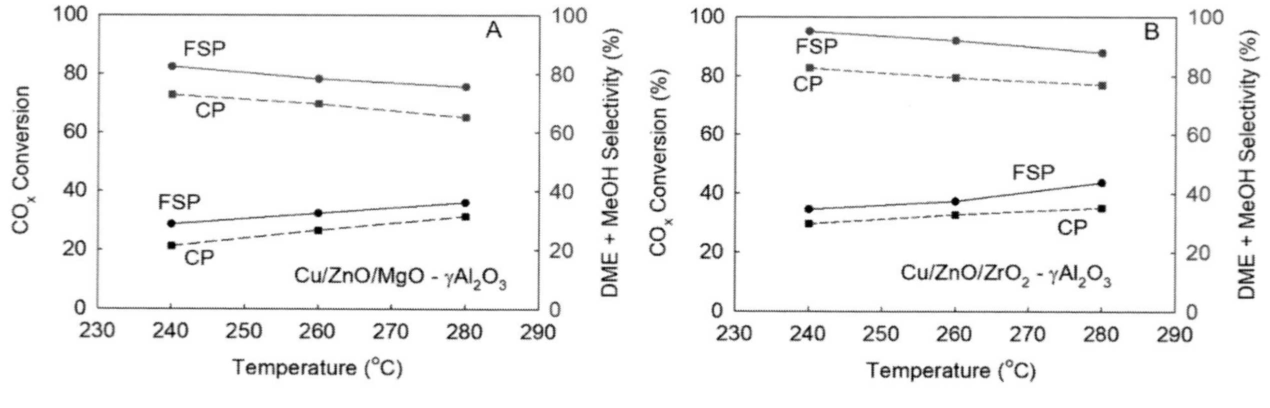

Effect of synthesis method on conversion of carbon oxides (CO + CO_2) and combined selectivity of Dimethyl ether (DME) and methanol (MeOH) on A) Cu/ZnO/MgO – γAl_2O_3 B) Cu/ZnO/ZrO_2 - γAl_2O_3. Flame synthesis (FSP), Co-precipitation (CP), Carbon oxides (CO_x).

84t A Modeling of Flow Cell eCO_2r System for Elucidating the Phenomena of Local Reaction Environment Using Multi-Physics Simulation.

Hyeonggeon Lee, Ung Lee

Many CO_2R studies have been carried out to suppress CO_2 emission in various fields. Among them, the eCO_2Rs have drawn attention due to the potential to be a low-carbon, sustainable way of producing a range of valuable chemicals. They have been developed intensively to break through their performance limitations and commercialize. Recently, the developed systems have consequently reached their industrial applicability. However, there is little understanding of how various properties of the system interactively affect the local reaction environment. Additionally, the existing mathematical models are limited in describing properly the intertwined phenomena in the system.

Motivated by these limitations, we developed a full-scale multi-physics model interpreting the eCO_2R. A vapor-fed flow-cell eCO_2R system is implemented using multi-physics simulation to analyze the role of EDL and cation effect on the local reaction environment. The experimental results show differences in current densities depending on the cations since the cations in electrolyte exhibit the Frumkin effect and steric effect, influencing the concentrations of ions and the potential in vicinity of cathode surface. These effects are addressed fidelitously in our model, and the simulation results, showing different concentration and potential profiles with cations, are consistent with the experimental results. This framework will pave the way for elucidating the intertwined phenomena in the eCO_2R system.

84u CO2 Upgradation By Methane Coupling on Metal Ion-Exchanged Zeolites

Sundar Raam Swaminathan, Venugopal Balashanmugam, Niket Kaisare

The coupling of CO_2 and methane to form acetic acid and other oxygenates provides an atom-efficient way to convert two greenhouse gases to a valuable chemical. Such a transformation consumes less energy compared to CO2 utilization via catalytic reduction [1]. However, one of the key challenges is that the reported yield of acetic acid through this reaction is very low. Therefore it is necessary to identify a suitable catalyst for this transformation. Wu et al. [2] have determined that Zn^{2+} in Zn-ZSM-5 can heterolytically activate methane into $[Zn\text{-}CH3]^+$ even at relatively mild temperatures. As a result, Zn-substituted zeolites have received much research attention in recent years [3].

While these catalysts have already been studied in batch conditions [3], optimizing under flow reactor conditions facilitates mechanistic studies and scale-up. In this work, various Zn-exchanged zeolites are considered and the effects of zeolitic support framework type (MFI, MOR) and acidity (different Si/Al), and metal loading of substituted metal (Zn) are studied at optimized reaction condition in a flow reactor. Computational studies by Zhang et al. [4] have identified that monovalent Zn^{2+} species located in the channel sites (straight and sinusoidal) of MFI zeolite exhibit the best catalytic activity [4]. While significant efforts have been put into post-synthetic modifications of ZSM-5, we have also synthesized ZSM-5 with selective Al-sites (to tune Al site population in either of the channels) suitable for effective Zn substitution in this study and their activity for CO_2 conversion will be analysed.

References

1. Álvarez ,A., et al., *ChemPhysChem*, **18** (2017) 3135–41.
2. Wu, J.-F., et al., J. Am. Chem. Soc., **135** (2013) 13567–73.
3. Shavi, R., et al., Appl. Catal. B: Environ., **229** (2018) 237–48.
4. Zhang, P., et al., Catal. Sci. Technol., **9** (2019) 6297–307.

84v A Structure-Guided Design of an Oligomeric Hydrophobin Bundle Using Coiled Coils

Mohamad Mahmoud, Suna Jo, Won Min Park

The class II hydrophobins, HFBI, are amphiphilic proteins secreted by filamentous fungi called *Trichoderma reesei*. These proteins are amphiphilic with a signature hydrophobic patch and can self-assemble at air-water or oil-water interfaces, forming highly ordered membrane-like structures. Due to their unique amphiphilicity property, hydrophobins have been utilized in various applications, such as emulsification, drug solubilization, protein purification, and functional materials fabrication. However, genetic modifications are restricted in their native fungal host, *T. reesei*. In contrast, the heterologous expression from yeast or bacterial hosts provides the benefits such as inexpensive nutrients requirement, rapid accumulation, and simpler industrial scale-up. However, forming the correct disulfide bonds buried in hydrophobins is challenging in bacterial expression systems. In addition, genetic modifications of hydrophobins for functionalization, without compromising their self-assembling property, are desirable but with limited success. We hypothesize that the genetic fusion of HFBI with an oligomeric protein motif whose shape matches the lattice geometry of self-assembled HFBI would direct the correct folding and protein self-assembly, improve protein expression, and enable functionalization. In this study, we designed fusion proteins of HFBI and a hexameric helical protein (HexCC), demonstrated their expression and purification from a bacterial host system, and characterized their interfacial assembly for functional coating materials fabrication.

We designed the recombinant fusion protein, HFBI-HexCC, by aligning HexCC with the hexagonal lattice of HFBI along their axes of symmetry. HexCC is a coiled-coil protein that forms a bundle-like hexameric oligomer. The sequences of HFBI and HexCC were fused using various unstructured linker sequences, followed by building comparative structure models from the crystal structure templates. From the results, we chose a linker sequence that showed the best geometric alignment without disrupting the overall structures of HFBI-HexCC. We performed the normal mode analysis to investigate the protein conformation and flexibility dynamics and confirmed the structural alignment between HFBI and HexCC. We also designed a fusion protein variant, $HFBI_{MC}$-HexCC, where the eight cysteine residues were mutated to alanine or valine to investigate the effects of coiled-coil fusion on the disulfide bond formation and protein folding. After the transformation of *Escherichia coli* cells with DNA plasmids, the fusion proteins were expressed, purified, and characterized. Both proteins were soluble after expression from *E. coli,* and the major fraction of proteins was in the form of hexamers or larger complexes. The fusion proteins were capable of self-assembling on the surface of polytetrafluoroethylene (PTFE) films, which was confirmed by the reduction in the surface hydrophobicity. Due to the protein layer formation on the surface PTFE, the water contact angle decreased from ~106° to ~73°. The mutants $HFBI_{MC}$-HexCC also showed a similar property. We also tested protein self-assembly on a polystyrene surface and investigated their potential as anti-fouling coating materials. Polystyrene microwell plates were coated with both proteins, and the adherence of *E. coli* cells was studied by quantifying the fluorescence of the green fluorescent protein, which is expressed inside the cells. The protein-coated polystyrene surfaces showed a twenty-fold reduction in the number of adhered cells compared to the bare control surfaces.

In conclusion, we demonstrated the design of oligomeric amphiphilic fusion proteins derived from fungal hydrophobin. Guided by the computational structural modeling, we fused HFBI with a coiled-coil, which resulted in a successful protein expression in a bacterial host system. In addition, the fusion proteins exhibited oligomer formation and self-assembling properties on polymer surfaces.

These fusion proteins can be used for biocompatible modification of various hydrophobic materials. Functionalization by further genetic modifications is possible, enabling them to serve as surface-binding agents with promising potential in biomedical, surface adhesion, and coatings applications.

84w Single Particle Cryo-EM Structure of Ferritin Biomineralization Showing the Protein-Nanoparticle Conjugate

Sagnik Sen

Ferritins are a class of ubiquitous, globular proteins that can assemble into a highly symmetric nanocage (1) (2). Ferritins mineralize iron nanoparticles within their inner hollow shell and are responsible for iron homeostasis in organisms (3). Comprehending the molecular mechanisms behind the mineralization will help better understand the fundamental processes involved and also can help further engineer the same for various bio-material synthesis applications. Single particle cryogenic electron microscopy (Cryo-EM) is a powerful structural biology tool which images proteins in their native functional state and has been previously used to obtain an atomic resolution model of Apoferritin (4). In our previous work (5), using Cryo-EM for studying two different biomineralization cases (including Ferritin) we obtained a low resolution map in case of Ferritin and thus not much structural information at the mineralization interface could be obtained. In this study we use Cryo-EM data processing including nanoparticle masking, to study the biomineral formation in Human Light Chain Ferritin (HuLF). We present a 4.4 Å structure of the Ferritin-Iron nanoparticle complex. This resolution allowed us to fit the model of Apoferritin (PDB 2FFX) to the electron density map we obtained. The amino acid residues at the protein-nanoparticle interface can be observed and provide information regarding interacting molecules responsible for the mineralization. Previous X- ray crystallography studies have shown glutamic acid residues on the surface of the inner light chain Ferritin shell to be responsible for the formation of the initial iron nucleation cluster (6) (7). Our structure shows the formed iron nanoparticle to be centered around the above mentioned residues.

Thus this sets up the grounds for the next part of our study to involve collection of larger data sets. Data collection with more particles having the protein-nanoparticle complex and preferably using energy filters would facilitate higher resolution reconstruction, subsequently providing an atomic level understanding of the entire process. This study further sets the template for analysis of more protein mediated biomineralization cases which can lead to better rational design of bioengineered nanomaterials.

References:

1. Honarmand Ebrahimi, Hagedoorn, P.-L., & Hagen, W. R. (2015). Unity in the Biochemistry of the Iron-Storage Proteins Ferritin and Bacterioferritin. *Chemical Reviews*, *115*(1), 295–326. https://doi.org/10.1021/cr5004908
2. Wang, Li, C., Ellenburg, M., Soistman, E., Ruble, J., Wright, B., Ho, J. X., & Carter, D. C. (2006). Structure of human ferritin L chain. Acta Crystallographica. Section D, Biological Crystallography., 62(7), 800–806. https://doi.org/10.1107/S0907444906018294
3. Zhang, Yu, X., Xie, J., & Xu, H. (2021). New Insights into the Role of Ferritin in Iron Homeostasis and Neurodegenerative Diseases. Molecular Neurobiology, 58(6), 2812–2823. https://doi.org/10.1007/s12035-020-02277-7
4. Irimia-Dominguez, Sun, C., Li, K., Muhoberac, B. B., Hallinan, G. I., Garringer, H. J., Ghetti, B., Jiang, W., & Vidal, R. (2020). Cryo-EM structures and functional characterization of homo- and heteropolymers of human ferritin variants. Scientific Reports, 10(1), 20666–20666. https://doi.org/10.1038/s41598-020-77717-4
5. Sen, Thaker, A., Sirajudeen, L., Williams, D., & Nannenga, B. L. (2022). Protein–Nanoparticle Complex Structure Determination by Cryo-Electron Microscopy. ACS Applied Bio Materials,

5(10), 4696–4700. https://doi.org/10.1021/acsabm.2c00130
6. Pozzi, Ciambellotti, S., Bernacchioni, C., Di Pisa, F., Mangani, S., & Turano, P. (2017). Chemistry at the protein–mineral interface in L-ferritin assists the assembly of a functional (μ3-oxo)Tris[(μ2-peroxo)] triiron(III) cluster. Proceedings of the National Academy of Sciences - PNAS, 114(10), 2580–2585. https://doi.org/10.1073/pnas.1614302114
7. Ciambellotti, Pozzi, C., Mangani, S., & Turano, P. (2020). Iron Biomineral Growth from the Initial Nucleation Seed in L-Ferritin. Chemistry : a European Journal, 26(26), 5770–5773. https://doi.org/10.1002/chem.202000064

84x Investigating the Potential of TMS-EDTA Modified Silica for Valuable Element Adsorption from Produced Water

Saeed Azizi, Anirban Ghosh, Songpei Xie, Mark Krzmarzick, Clint Aichele

This study investigated the potential of TMS-EDTA modified silica as an adsorbent for removing rare earth elements from produced water. Different pH values were used during the modification process, leading to the formation of TMS-EDTA on the surface of the silica. The effect of pH on the adsorption capacity of the modified silica was examined using high salinity mixtures solution containing lanthanum. FTIR was employed to characterize the modified silica, confirming the grafting of TMS-EDTA onto the surface of the silica powders. The results of zeta potential analysis demonstrated that a higher pH value during the modification process resulted in a lower negative charge on the surface of the silica, thus enhancing its capability to adsorb positively-charged ions. The modified silica with an optimum pH value of the modification process exhibited superior adsorption capacity compared to unmodified silica, exhibiting higher selectivity towards lanthanum, even in the presence of a high concentration of sodium. These findings suggest that TMS-EDTA modified silica possesses considerable potential for removing valuable components from wastewater with high salinities. The results also contribute to the growing body of research on effective strategies for the management of produced water, particularly in the context of environmental protection and resource recovery.

Treatment of Domestic Wastewater by Coagulation, Adsorption, and Filtration for Reusing in the Production of Concrete Mixtures

R. Morsy[1,2], Salma O. Mahmoud[1], Emad El-Din E. Abdel-Qader[1], Kareem M. Bakr[1], Anas K. Taha[1], Ahmed M. Abdallah[1], Hassan M. Abdul Karim[1], Ahmed S. Mahmoud[3], and Mohamed K. Mostafa[1]

[1] Faculty of Engineering and Technology, Badr University in Cairo (BUC), Cairo, Egypt

[2] *Housing and Building National Research Center (HBRC), Egypt*

[3] Institute of Environmental Studies, Arish University, North Sinai, Egypt

Abstract

Concrete production consumes huge amounts of fresh water for hydration of cement, concrete curing and washing aggregate, so providing an alternative is important to save fresh water. In this research, treated domestic wastewater is used for concrete production. The wastewater samples were collected from Badr University wastewater treatment plant in Cairo, Egypt. The study aims to show the effect of using treated wastewater to the concrete mixture on its mechanical properties and compare the results with control concrete mixture using fresh water. The experimental plan tests concrete with grade 50 MPa using: primary, secondary, tertiary treated wastewater, and fresh water as mixing water. The concrete mechanical properties: compressive, tensile splitting, and flexure strength at ages of 7, and 28 days are evaluated. The results of compressive strength test of cubes using primary and secondary treatment are almost similar to the results of the control mixture in the meanwhile the compressive strength of mixtures using tertiary treated wastewater reached the same compressive strength that of the control mixture. The effect of treated wastewater on concrete split tensile and flexure strength results are comparable to that of control mixture results. The results of the tertiary treated wastewater showed improved performance for concrete mixing, and it can be properly utilized for concrete production.

Keywords: Adsorption, Coagulation, Filtration, concrete mixtures, water reuse, Wastewater treatment

1. Introduction

Water is an important ingredient of concrete and affect concrete production that includes mixing, hydration of cement, washing aggregate, workability and curing for concrete. The quality of water must be carefully examined to avoid structural defects such as the presence of chemical impurities in water, that affect the strength and durability of concrete [1]. ASTM C1602 [2] permits the use of potable water without any testing however the non-potable water sources requires testing for qualifying non-potable water sources taking into account that it could affect the mechanical properties of concrete and it is crucial because impurities in it might prevent cement from setting properly, affect the strength of concrete, and may lead to reinforcement corrosion inside concrete. So, it is very important to quantify the quality of water for mixing and curing should be taken into account [3].

One of the method for saving potable water and avoid shortening of water is through treating and reusing the domestic wastewater [4]. There are several types of wastewater that can be introduced into the municipal system, such as household, hospital food industry, metal industry, and painting industry wastewater, and each of them requires a specific treatment process. There are several treatments suitable for wastewater including; physical, biological, and chemical approaches [5]. Coagulation and adsorption are a chemical treatment methods that draw the attention of many researchers [6-16].

The significance of this study is to find a potential for using treated wastewater instead of potable water for concrete production taking into consideration the code requirement for non-potable water usage in concrete. The study includes experimental tests for concrete workability, compressive strength, splitting tensile strength, and flexure strength of concrete using three different treatment stages of wastewater and using the potable water is used as a control mixture to compare the results with.

2. Material and Methods

Concrete mixtures prepared in the laboratory are made using the following materials: cement, gravel, sand, cement, and treated wastewater with different treatment stages. The experimental study includes twelve different mixtures testing concrete with grade 50 MPa using: primary, secondary, tertiary treated wastewater, and potable water as mixing water. The cement used in this study is normal Portland cement N42.4, where its properties are conforming to ASTM C150-07. The Blaine fineness is 3154.5 cm^2/gm and specific gravity is 3.13. The coarse aggregates used in the concrete mixture are a crushed 20 mm size. The coarse aggregate used, its specific gravity is 2.65, and absorption capacity equals 1.95%. Natural sand (5 mm) maximum size, and with 0.9% passing the 75 μm (No. 200 ASTM) sieve was used. Its specific gravity is 2.5, and its SSD condition water absorption of 0.98%. A water test was carried out in and its parameters studied were physical properties (TDS (mg/l), TSS (mg/l), Tp (mg/l)) and chemical properties (pH, Chloride (mg/l), given in Table 1.

Table 1 Water Test for wastewater with different treatment stages

Test	Before treatment	Stage 1	Stage 2	Stage 3
Ph.	5	6.5	7.2	7.8
CL	800	590	410	300
TDS	3500	3000	2200	1650
TSS	730	700	660	310
Tp	22	17	9	3.1

3. Experimental Testing Plan

3.1 Mixture design

In the current study, twelve mixtures were performed made with different mixing and curing water and a reference mixture using potable water. The produced concrete series are coded according to the primary, secondary, tertiary treated wastewater, and fresh water as mixing water in the concrete mix and using water-cement ratio equals to 0.45, as given in Table 2.

Table 2: Concrete mixture description

Mixtures	Cement	Fine aggregates	Coarse aggregates	Mixing water	Curing
M0	475	522	1218	PW	PW
M10	475	522	1218	PTW	PW
M11	475	522	1218	PTW	PTW
M12	475	522	1218	PTW	STW
M13	475	522	1218	PTW	TTW
M20	475	522	1218	STW	PW
M21	475	522	1218	STW	PTW
M22	475	522	1218	STW	STW
M23	475	522	1218	STW	TTW
M30	475	522	1218	TTW	PW
M31	475	522	1218	TTW	PTW
M32	475	522	1218	TTW	STW
M33	475	522	1218	TTW	TTW

3.2 Slump Test

Concrete mixing is done in a laboratory using pan mixer and slump tests were performed on the mixes to measure consistency as described in ASTM C1611 / C1611M [17]. The mixes were then prepared and cured according to ASTMC192/192M-06 [18]. Concrete cubes with dimensions 150x150x150 mm and concrete cylinders with dimensions 150x 300 mm were used for testing. The specimens were mixed and compacted on a vibrating compactor and the curing where all specimens were immersed in water tank. The results for the slump test have the same values 65 mm for the whole testing mixtures.

3.3 Compressive Strength Test

Compressive strength tests were conducted on concrete cubes of size 150x150x150 mm according to BS EN 12390-3:2019 [19] using ELE international ADR auto v2.0 3000 BS EN Compression Machine. The concrete cubes were tested at 7, 14, and 28 days.

3.4 Splitting Tensile Strength Test

A Split tensile strength test on the concrete cylinder size 150x300 mm is a method to determine the concrete's tensile strength. The procedure is based on the ASTM C496 / C496M-17

Standard Test [20], using ELE international ADR auto v2.0 3000 BS EN Compression Machine.

3.5 Flexure Strength Test

Flexural strength tests for plain concrete beam failure due to bending. In the current study, beams of 150 mm × 150 mm × 700 mm were tested under two-point loading as the span between the columns is small. At this flexural strength, the effective length of the beam was given to be 640 mm. The specimens were tested after 28 days using the Universal testing machine 600KN as per ASTM C78/C78M-22 [21].

4. Result and discussions

4.1 Compressive Strength Test

Figure 1 presents the compressive strength results of designed mixtures. It is observed from the results, that the compressive strength for the specimens withing TTW as a mixing water and also using TTW for concrete specimens curing gives a comparable values to that for the specimen mixed and cured using potable water. The concrete compressive strength values decreased using the STW, and PTW as a mixing water and cured water than the compressive strength values using the TTW and PW. For the specimens cured by PW, as shown in Figure 1 (a) the compressive strength values for concrete cubes at 28 days using TTW as a mixing water decreased by 3.8% of the values using PW. The compressive strength values of the concrete cubes mixed with STW decreased by the 6.7% of the control mixture mixed with PW, the decrease of the compressive strength cubes mixed with PTW reached to 8.7% compared to the control mixture. Using PTW as a mixing water, as shown in Figure 1(b), the decrease in the compressive strength at 28 days for the specimens cured using PTW compared to the control mixture equals to 17.3%, the decrease reached to 11.5% for the specimen cured with STW. However, the decreased in the compressive strength at 28 days equals to 9.6% for the cubes cured with TTW. For the specimens mixed with STW, as shown in Figure 1(c), the decrease in the compressive strength at 28 days for the specimens cured using PTW compared to the control mixture equals to 15%, the decrease reached to 13.5% for the specimen cured with STW. However, the decreased in the compressive strength at 28 days equals to 8.4% for the cubes cured with TTW. And for concrete specimens mixed with TTW, as shown in Figure 1(d), the decrease in the compressive strength at 28 days for the specimens cured using PTW compared to the control mixture equals to 12.4%, the decrease reached to 9.3% for the specimen cured with STW. However, the decreased in the compressive strength at 28 days equals to 5.4% for the cubes cured with TTW.

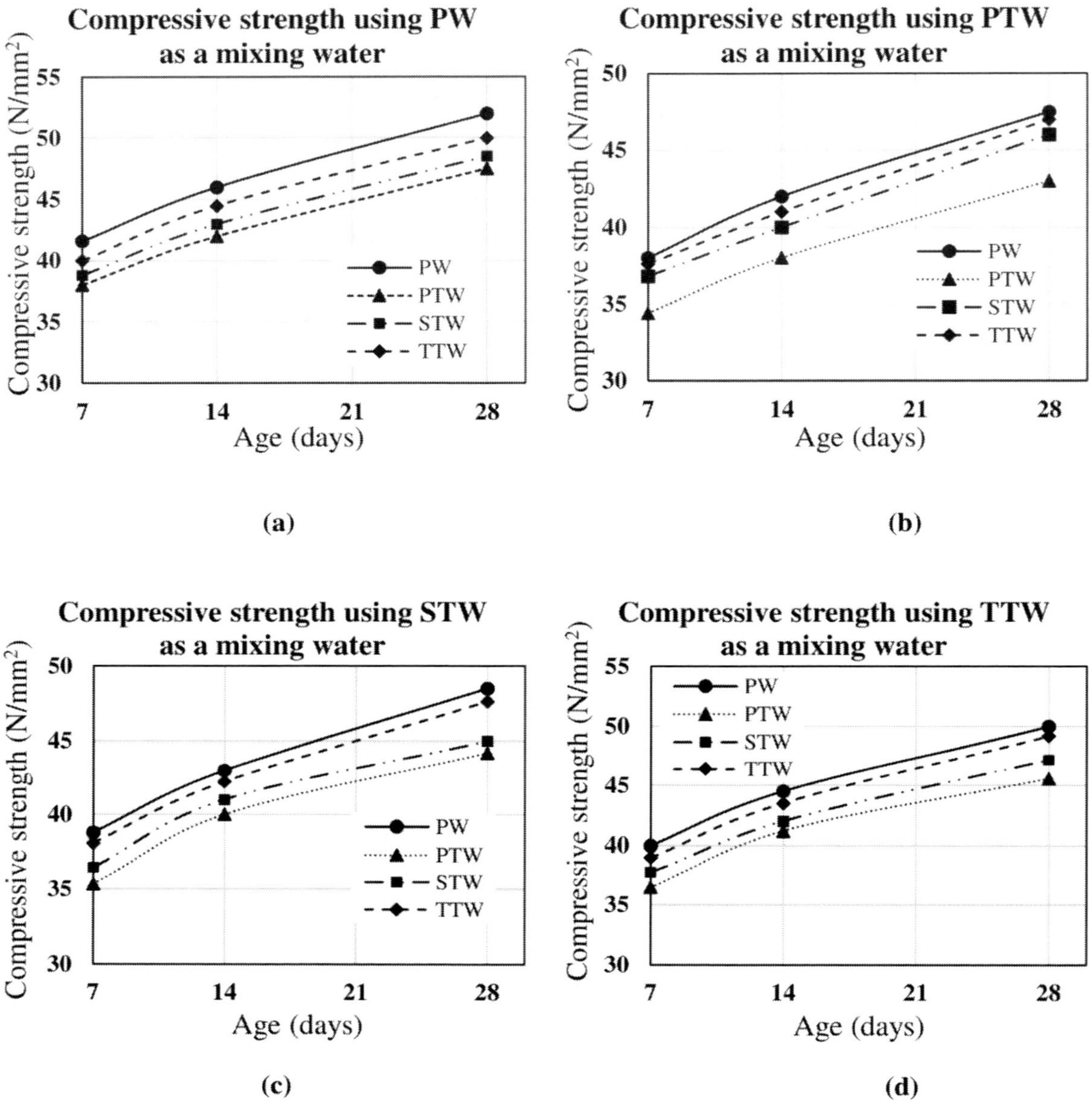

Figure 1 Compressive strength test results for the mixtures

4.2 Tensile Splitting Strength Test

Concrete cylinders of 28 days curing period were tested for splitting tensile strength evaluating using different stages for treated wastewater. Three cylinders samples of each mixture were tested. The average strength of these three samples was taken as the result. The results of flexural strength are shown in Figure 2, the results are categorized according to the water used for curing tested specimens. The results are obviously showing the influence of using treated wastewater on the tensile splitting strength. For specimens cured by PW, the decrease in the

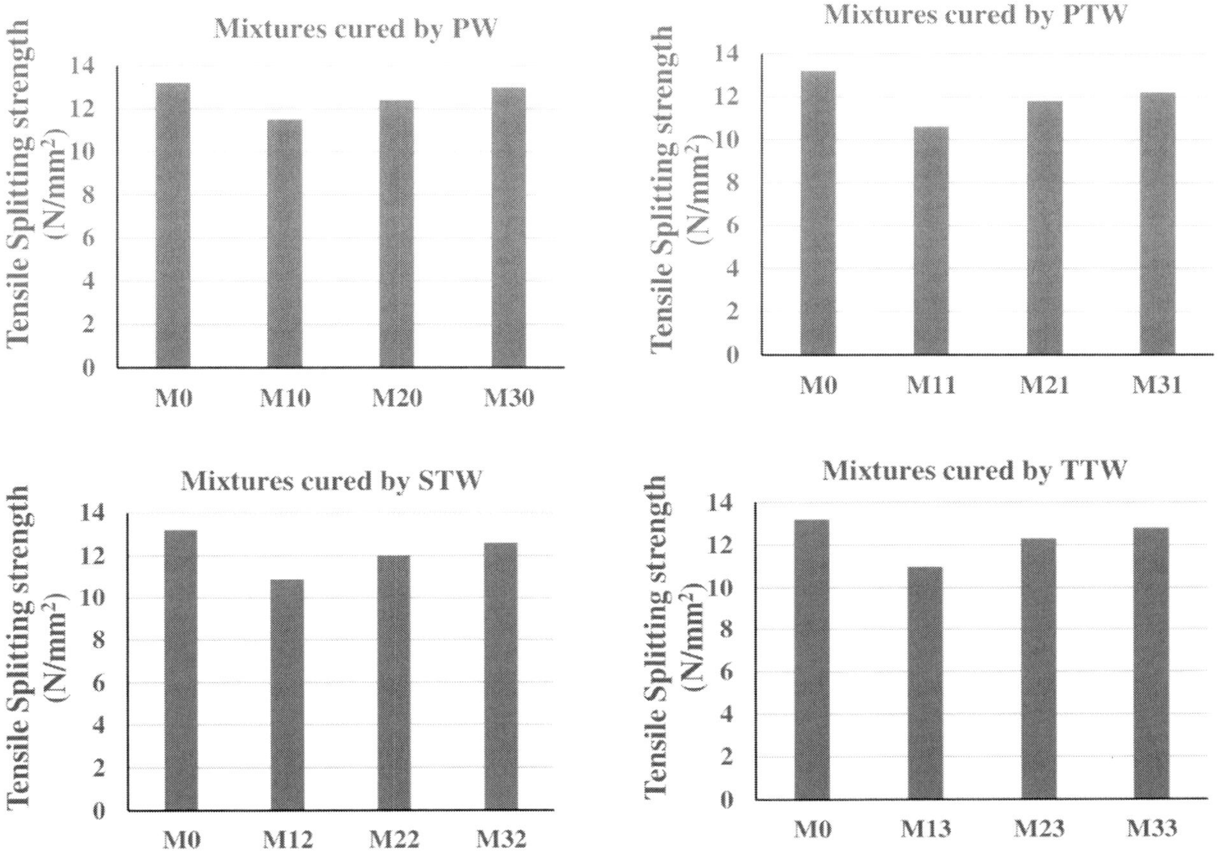

Figure 2 Tensile splitting strength test results for the mixtures

tensile strength reached to 8.7% with using PTW for mixing, while the decease in the strength equals to 6.7% for the cylinders mixed with STW. However, the decreased in the tensile splitting strength at 28 days equals to 3.8% for the cylinders mixed with TTW.

4.3 Flexure Strength Test

Concrete specimens prisms of 28 days curing period were tested for flexural strength using different stages for treated waste water. Three samples of each mixture were tested. The average strength of these three samples was taken as the final result. The results of flexural strength are shown in Figure 3. From the Figures, it is found that the flexure strength for the control mixture equals 15 MPa, while flexure strength showed a decrease up to 7.5 % for specimen mixed with PTW and cured with PW, the decrease reached to 14.5% for the specimens mixed with PTW and cured with PTW, a 11.2% reduction in the flexure strength for prisms mixed with PTW and cured with STW and a 9.2% reduction in strength for specimens mixed with PTW and cured by TTW.

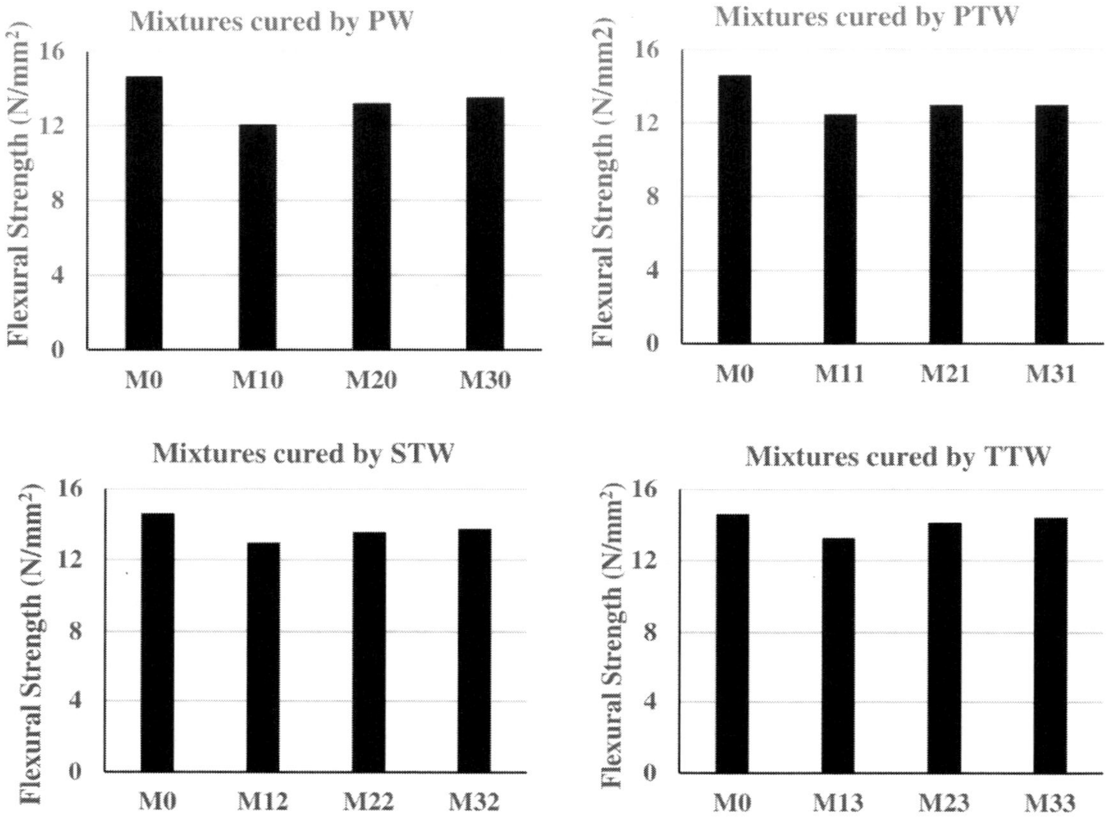

Figure 3 Flexure strength test results for the mixtures

5. Conclusions

It is concluded that it is possible to replace treated wastewater instead of potable water in concrete mixtures starting from the first stage due to the low cost and quality not much less than potable water and the possibility of using it with the assurance that it will perform the same purpose with the same efficiency and this will also leads to save a lot of potable water consumption and benefit from it in many other fields and to preserve the environment through the use of wastewater and avoid the damages and diseases that it causes and it will lead to our success as civil engineers from our role in the water conservation plan.

Acknowledgements

The authors gratefully acknowledge the financial support given by the Joint Academy of Scientific Research and Technology (ASRT)/the Bibliotheca Alexandrina (BA) Research Grants (Grant Number: 1469).

References

[1] N.R. CHANDNE, A.V. SHIRGIRE. 2021. Effect of different sources of water on strength of concrete. International Journal of Research Science and Management. 4,2, 111-114.

[2] ASTM C1602/C1602M-06, Standard Specification for MiXing Water Used in the Production of Hydraulic Cement Concrete. (C),1-5.

[3] P.P. YALLEY, K. APPIADU-BOAKYE, W.D. ADZRAKU. 2019. Assessing qualities of different sources of water for mixing concrete, Journal of Engineering Research and Reports. 5, 4, 1–8.

[4] EL-SHAFEI, M., MAHMOUD, A., MOSTAFA, M. & PETERS, R. Effects of entrapped nZVI in alginate polymer on BTEX removal. AIChE Annual Meeting. San Francisco, CA, 2016. 13-18.

[5] MAHMOUD, M. & MAHMOUD, A. S. 2021. Wastewater treatment using nano bimetallic iron/copper, adsorption isotherm, kinetic studies, and artificial intelligence neural networks. Emergent Materials, 4, 1455-1463.

[6] Farag, R. S., Elshfai, M. M., Mahmoud, A. S., Mostafa, M. K. & Peters, R. W. (592d) Green Synthesis of Nano Iron Carbide: Preparation, Characterization and Application for Removal of Phosphate from Aqueous Solutions. Annual AIChE Meeting Proceedings; Pittsburgh, PA, October 28 - November 2 https://www.aiche.org/conferences/aiche-annual-meeting/2018/proceeding/paper/592d-green-synthesis-nano-iron-carbide-preparation-characterization-and-application-removal (2018).

[7] Heberling, J., Adewuyi, Y., Mahmoud, A. S., Mostafa, M. & Peters, R. W. (12e) AOP Performance at Wastewater Treatment Plants | AIChE. AIChE Annual Meeting Proceedings, David L. Lawrence Convention Center, Pittsburgh, PA, October 28 - November 2 https://www.aiche.org/conferences/aiche-annual-meeting/2018/proceeding/paper/12e-aop-performance-wastewater-treatment-plants (2018).

[8] Mahmoud, A. S., Mostafa, M. K. & Peters, R. W. A prototype of textile wastewater treatment using coagulation and adsorption by Fe/Cu nanoparticles: Techno-economic and scaling-up studies. Nanomater. Nanotechnol. 11, (2021).

[9] Mahmoud, A. S., Mohamed, N. Y., Mostafa, M. K. & Mahmoud, M. S. Effective Chromium Adsorption From Aqueous Solutions and Tannery Wastewater Using Bimetallic Fe/Cu

Nanoparticles: Response Surface Methodology and Artificial Neural Network. Air, Soil Water Res. 14, (2021).

[10] Mahmoud, M. S., Mahmoud, A. S., El-Said, M. A. & Mostafa, M. K. Comparison of aluminum and iron nanoparticles for chromium removal from aqueous solutions and tannery wastewater, empirical modeling and prediction. Emergent Mater. 5, 1729–1744 (2022).

[11] Mahmoud, A. S., Mahmoud, M. S., Yousri, N. & Mostafa, M. K. Advanced Wastewater Treatment by Using Entrapped nZVI into Alginate (Ag) Biopolymer: Adsorption Isotherm, Kinetic Models, and Statistical Analysis. Key Eng. Mater. 921, 147–160 (2022).

[12] Mahmoud, A. S., SaryEl-deen, R. A., Mostafa, M. K. & Peters, R. W. (655d) Artificial Intelligence for Organochlorine Pesticides Removal from Aqueous Solutions Using Entrapped Nzvi in Alginate Biopolymer | AIChE. AIChE Annual Meeting Proceedings https://www.aiche.org/conferences/aiche-annual-meeting/2017/proceeding/paper/655d-artificial-intelligence-organochlorine-pesticides-removal-aqueous-solutions-using-entrapped (2017).

[13] Mostafa, M. K., Mahmoud, A. S., Mahmoud, M. S. & Nasr, M. Computational-Based Approaches for Predicting Biochemical Oxygen Demand (BOD) Removal in Adsorption Process. Adsorpt. Sci. Technol. 2022, (2022).

[14] El-Shafei, M. M., Mahmoud, A. S., Mostafa, M. K. & Peters, R. W. Effects of entrapped nZVI in alginate polymer on BTEX removal. Environmental Division 2016 - Core Programming Area at the 2016 AIChE Annual Meeting 15–22 https://www.aiche.org/conferences/aiche-annual-meeting/2016/proceeding/paper/60b-effects-entrapped-nzvi-alginate-polymer-on-btex-removal (2016).

[15] Farag, R. S. et al. (670d) Study the degradation and adsorption processes of organic matters from domestic wastewater using chemically prepared and green synthesized nano zero-valent iron. AIChE Annual Meeting, Conference Proceedings vols 2019-Novem https://www.aiche.org/conferences/aiche-annual-meeting/2019/proceeding/paper/670d-study-degradation-and-adsorption-processes-organic-matters-domestic-wastewater-using-chemically (2019).

[16] Kamar, M. T., Elattar, H., Mahmoud, A. S., Peters, R. W. & Mostafa, M. K. A critical review of state-of-the-art technologies for electroplating wastewater treatment. Int. J. Environ. Anal. Chem. (2022) doi:10.1080/03067319.2022.2098486.

[17] ASTM C143/C143M-2020, Standard Test Method for Slump of Hydraulic-Cement Concrete.

[18] ASTM C192/C192M-2019, Standard Practice for Making and Curing Concrete Test Specimens in the Laboratory.

[19] BS EN 12390-3. 2019. Testing hardened concrete- Part 3: Compressive Strength of Test Specimens. British Standards Institute (BSI).

[20] ASTM C496/C496M-04e1, Standard Test Method for Splitting Tensile Strength of Cylindrical Concrete Specimens.

[21] ASTM C78/C78M-22, Standard Test Method for Flexural Strength of Concrete (Using Simple Beam with Third-Point Loading).

84z High-Efficiency Solar Evaporator System for Treating Brackish and Produced Water.

Mounika Chevula, Punya Mainali, Charlie Cutts, Sharad Puri, David N. McIlroy, Seokjhin Kim

A high-efficiency solar evaporator designed to address the growing need for energy-efficient and innovative desalination systems. The solar evaporator is developed for treating brackish and produced water (BPW) in regions where water scarcity and quality issues are prevalent, utilizing cost-effective and reliable solar energy. The system features a high surface area coating of 1D nanostructures with a solar collector, integrated into a simple bulb housing with black-painted metal parts to maximize infrared absorption and preheat the water before entering the evaporation zone. To increase the evaporation rate of the solar evaporator, the surface is coated with a layer of unique 1D nanostructures called silica nano springs. The nano springs were prepared using the vapor-liquid-solid mechanism, and a thin layer of carbon was conformally coated on their surface using atmospheric pressure chemical vapor deposition. The study demonstrates that incorporating carbon-coated silica nano springs significantly improves the efficiency of the solar-driven desalination system, making it a simple and cost-effective method for effectively desalinating BPW for irrigation purposes. The use of nanomaterials in solar-powered desalination systems highlights their potential for improving efficiency and cost-effectiveness in water treatment.

84aa Demonstration of a Batch Electrochemical System for Phosphorus Recovery from a Real Municipal Wastewater Recycle Stream

Syed Asad Abbas, Sana Heydarian, Lawrence Ajayi, Jason Trembly, Damilola Daramola

Research Interests: 1) Renewable Energy

2) Energy Conversion

3) Water Splitting

4) Water Treatment

Phosphorus is an essential nutrient for both animals and plants, but it has an adverse effect on the ecosystems of water streams due to eutrophication [1]. Phosphorus recovery from wastewater streams is an essential part of wastewater treatment plants to curb the destruction of ecosystems in downstream water bodies. The idea of recirculation of phosphorus from wastewater treatment plants to grain fields will not only provide a continuous fertilizer supply but it will also minimize the scarcity issues associated with phosphorus in the world of growing phosphorus demand [2]. Sustainable process of the phosphorus recovery in the form of fertilizer from wastewater treatment plants requires optimal reactor design, high recovery rates, and minimum energy consumption [3]. Electrochemical systems have an additional advantage of being utilized using renewable energy resources. In this study we have established a rational design for the sustainable removal of phosphorus using an electrochemical system in the form of struvite or brushite, which are magnesium-based and calcium-based phosphates, respectively [4]. A complete range of experiments were performed for the investigation of phosphorus recovery efficiency and the specific energy consumption during recovery. Chronoamperometry, Cyclic voltammetry and Impedance analysis were performed for a real wastewater sample to investigate the reaction mechanism, and energy consumptions. Liquid Chromatography (LC) and Inductively Coupled Plasma (ICP) were utilized to quantify the recovery rates in the liquid streams. Scanning Electron Microscopy (SEM), X-ray Diffraction (XRD) and Raman analysis were performed for quantitative and qualitative analysis of recovered solid products. These results will be presented at the upcoming conference.

References.

- Wang Yingmu, Ziyuan Lin, Yue Wang, Wei Huang, Jiale Wang, Jian Zhou, and Qiang He. Sulfur and Iron Cycles Promoted Nitrogen and Phosphorus Removal in Electrochemically Assisted Vertical Flow Constructed Wetland Treating Wastewater Treatment Plant Effluent with Jigh S/N Ratio. *Water research 151, (2019), 20-30.*
- Devos Perrine, Ahlem Filali, Paloma Grau, and Sylvie Gillot. Sidestream Characteristics in Water Resource Recovery Facilities: A Critical Review. *Water Research 232, (2023) 119620.*
- Pindine Garrett P., Jason P. Trembly, and Damilola A. Daramola. Equilibrium-Based Temperature-Dependent Economic Analysis of the Recovery of Phosphorus from Different Wastewater Streams via Chemical Precipitation. *ACS ES&T Water 1, No. 11 (2021) 2318-2326.*
- Belarb, Zineb, Damilola A. Daramola, and Jason P. Trembly. Bench-Scale Demonstration and Thermodynamic Simulations of Electrochemical Nutrient Reduction in Wastewater Via Recovery as Struvite. *Journal of The Electrochemical Society 167, No. 15 (2020): 155524.*

84ac Modified Donnan Dialysis Process for Selective Nutrient Removal from Agricultural Liquid Waste

Amir Akbari, Lauren F. Greenlee, Bruce E. Logan

Significant amounts of important macronutrients such as phosphorus, nitrogen, and potassium in high-strength wastewaters such as animal manures, makes nutrient recovery from such waste streams a compelling opportunity to achieve a circular economy and address increasing global fertilizer demand. However, the proposed technologies should be able to deal with high suspended solids, organic matter content, and ion concentrations of such waste streams. Donnan dialysis with ion exchange membranes (IEMs), are one technology option with low energy consumption that has the potential to deal with these challenges and could be used for nutrient recovery from different wastewater sources.

In this talk, I will discuss a new modified cation exchange membrane (CEM)-separated Donnan dialysis process in which a sacrificial magnesium plate is placed in the anolyte to recover phosphate precipitates through electrochemically induced reactions. In addition, the cell produces a side-stream of liquid fertilizer for fertigation applications through the simultaneous separation of ammonium and potassium. A simulated agricultural liquid waste, based on water chemistry of real animal manure samples, has been used as the anolyte. Two different cation exchange membranes have been tested to selectively separate ammonium and potassium from wastewater, while the pure sacrificial magnesium plate in the anolyte was the only source of magnesium in the system. Also, different draw solutions have been tested as the catholyte to understand the role of draw ions on overall system performance.

Inductively coupled plasma atomic emission spectroscopy (ICP-AES) was used to track ion concentrations in each chamber over time for kinetic purposes. A set of characterization techniques including scanning electron microscopy (SEM), X-ray diffraction (XRD), and Fourier-transform infrared spectrometry (FT-IR) were used to identify the morphology and nature of the precipitates. The results indicated complete inorganic phosphorus removal in form of solid precipitates, mainly struvite. Also, selective separation between ammonium and potassium was achieved with decreased ammonium ion cross over the CEM compared to potassium. These results show that the proposed process can successfully recover valuable, diverse products from wastewater while avoiding the disadvantages of other commonly used chemical precipitation methods of struvite recovery, such as magnesium salt dosing and adding base to the system for pH control.

84ad A Stochastic Optimization and Machine Learning-Based Framework for Evaluating Ammonia Utilization As a Hydrogen Carrier

Dongjun Lim, Jiwon Gu, Jong Ah Moon, Yeong Jin Koh, Hankwon Lim

Ammonia is a highly versatile chemical compound consisting of hydrogen and nitrogen. It has been widely used in various conventional industries such as agriculture, refrigeration, and cleaning agents. However, one of the most promising applications of ammonia is its use as a hydrogen carrier due to its relatively high hydrogen content of 17.7% by weight.[1] Using ammonia as a hydrogen carrier provides a safer and more efficient way to store and transport hydrogen compared to using compressed hydrogen gas that requires high pressure (>350 bar) and has some inherent risks such as high flammability and low volumetric energy density.[2, 3]

In addition, ammonia can also be used as a feedstock for producing hydrogen through ammonia decomposition. During ammonia decomposition, hydrogen and nitrogen gas are produced using various methods including catalytic or thermal decomposition. Compared to other hydrogen production methods like steam reforming of methane, ammonia decomposition requires lower energy input making it a promising option for reducing carbon dioxide emissions.[4] Overall, the versatility of ammonia as a hydrogen carrier and feedstock for hydrogen production makes it an attractive candidate for sustainable energy applications. Ongoing research efforts are exploring the potential of ammonia in various areas, such as fuel cells, transportation, and power generation. Advancements in this area may lead to significant breakthroughs in the transition towards a more sustainable and cleaner energy future.

Despite its potential as a hydrogen carrier and feedstock for hydrogen production, the current ammonia production process, the Haber-Bosch process, has some significant drawbacks. The process requires massive amounts of energy and emits enormous carbon dioxide because the Haber-Bosch process involves reacting hydrogen from fossil fuels with nitrogen from the air at high temperatures and pressure.[5] Additionally, the high energy requirements of the process result in high production costs.

Therefore, the overall costs and emissions associated with ammonia utilization as a hydrogen carrier need to be carefully evaluated to ensure that it is a sustainable and cost-effective solution. While ammonia has several advantages over compressed hydrogen gas as a hydrogen carrier, it also has its own production costs and emissions associated with it. The hydrogen used in the ammonia production process comes from fossil fuels, so the cost of producing ammonia inherently includes the cost of producing hydrogen. Thus, it is essential to consider the life-cycle cost and emissions associated with the entire process, including the production, transportation, and use of ammonia as a hydrogen carrier. By evaluating these factors, we can ensure that ammonia utilization is a sustainable and cost-effective solution for transitioning towards a cleaner and more sustainable energy future.

This study aims to investigate the feasibility of utilizing ammonia as a hydrogen carrier and to propose an efficient evaluation framework for the overall cost and emissions associated with ammonia utilization strategies such as replacement of fossil fuel in on-site hydrogen refueling station or fuel cell power plants. Specifically, we develop an optimization framework that leverages the Monte Carlo method and machine learning to import ammonia into the Republic of Korea. To achieve this, we will begin by selecting several exporting countries based on historical import-export data and developing predictive models for ammonia production costs and associated emissions. With this information, we will establish trading routes between exporting countries and KOR taking into account the location of actual ammonia manufacturing complexes and export terminals. Additionally, we will construct a supply chain from KOR terminals to the actual ammonia demand.

To evaluate the entire cycle of ammonia, we will use a techno-economic analysis and optimize it in terms of life-cycle cost and carbon dioxide emissions using the Monte Carlo method. Based on the stochastic optimization results, we will develop several machine learning models and select the most suitable predictive model to make informed decisions regarding ammonia utilization. Our findings will provide insights into the feasibility of using ammonia as a hydrogen carrier and the potential benefits of using it to reduce carbon dioxide emissions.

With the stochastic optimization method, multiple scenarios can be established for ammonia utilization strategies such as determining the countries involved in trading and the method of ammonia production. Additionally, a system that users want to investigate can be optimized to minimize overall cost and environmental impacts. By utilizing machine learning-based regression models trained with the results of stochastic optimization, users can input specific parameters and obtain the total ammonia cost, unit product price (e.g., USD kWh^{-1} or USD kgH_2^{-1}), and life-cycle emissions efficiently. Therefore, this study enables to efficiently analyze the feasibility of ammonia-based hydrogen facilities. In conclusion, this study provides a framework for efficiently evaluating the cost and emissions of different ammonia utilization strategies through optimizing ammonia import and training machine learning models for regression.

Reference

1. Li Y, Bi M, Li B, Zhou Y, Huang L, Gao W. Explosion hazard evaluation of renewable hydrogen/ammonia/air fuels. *Energy*. 2018;159:252-263.

2. Hua TQ, Ahluwalia RK, Peng JK, et al. Technical assessment of compressed hydrogen storage tank systems for automotive applications. *Int J Hydrogen Energy*. 2011;36(4):3037-3049. doi:10.1016/J.IJHYDENE.2010.11.090

3. Eberle U, Arnold G, Von Helmolt R. Hydrogen storage in metal–hydrogen systems and their derivatives. *J Power Sources*. 2006;154(2):456-460.

4. Cha J, Park Y, Brigljević B, et al. An efficient process for sustainable and scalable hydrogen production from green ammonia. *Renew Sustain Energy Rev*. 2021;152:111562.

5. Zhang L, Li R, Cui L, et al. Boosting photocatalytic ammonia synthesis performance over OVs-Rich Ru/W18O49: Insights into the roles of oxygen vacancies in enhanced hydrogen spillover effect. *Chem Eng J*. 2023;461:141892. doi:10.1016/J.CEJ.2023.141892

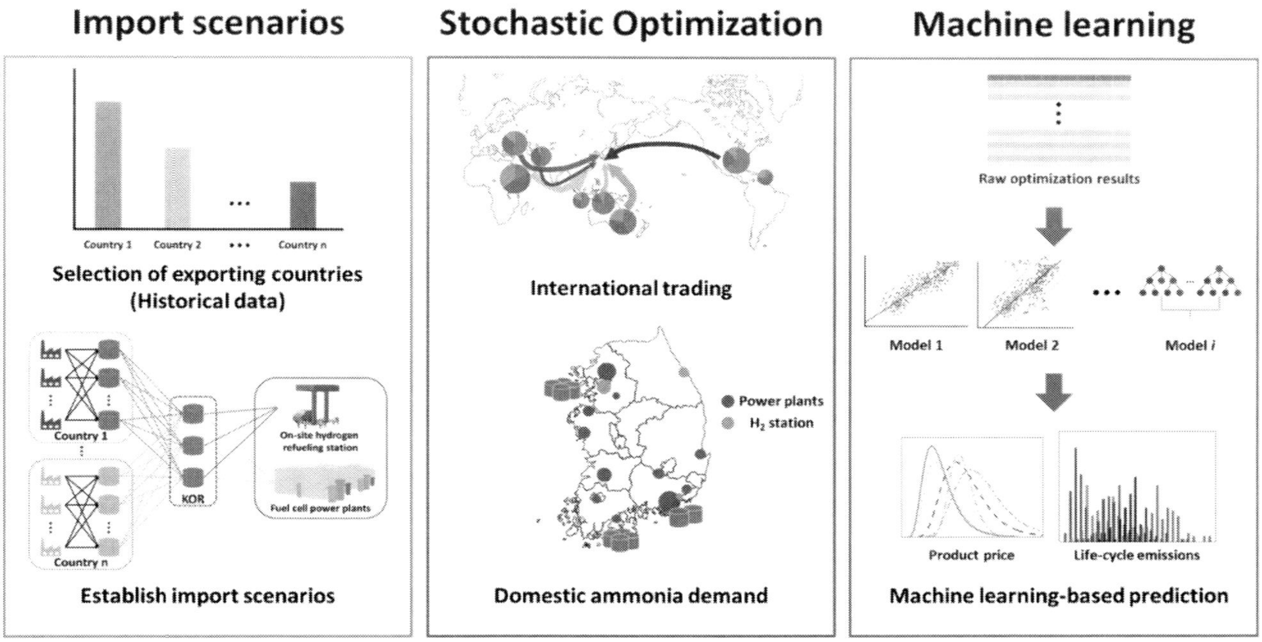

Import scenarios
Stochastic Optimization
Machine learning
Country 1
Country 2
Country n
Selection of exporting countries
(Historical data)
Country 1
Country n
KOR
On-site hydrogen refueling station
Fuel cell power plants
Establish import scenarios
International trading
Power plants
H_2 station
Domestic ammonia demand
Raw optimization results
Model 1
Model 2
Model i
Product price
Life-cycle emissions
Machine learning-based prediction

84af Developmental Pb Exposure Increases AD Risk Via Altered Intracellular Ca^{2+} Homeostasis in hiPSC-Derived Cortical Neurons

Junkai Xie, Shichen Wu, Chongli Yuan

Background: Exposure to environmental chemicals such as lead (Pb) during vulnerable developmental periods can result in adverse health outcomes later in life. Human cohort studies have demonstrated a strong association between developmental Pb exposure and Alzheimer's Disease (AD) onset in later life which were further corroborated by findings from animal studies. The molecular pathway linking developmental Pb exposure and increased AD risk, however, remains elusive.

Objective: In this work, we used human iPSC-derived cortical neurons as a model system to study the effects of Pb exposure on developing and maturing human cortical neurons.

Methods: Neural progenitor cells induced from human iPSC were exposed to 0, 15 and 50 ppb of lead which were further differentiated into mature cortical neurons. RNA-sequencing and immunofluorescence were used to study the changes in transcriptome and protein expression. ELISA assays for β-amyloid and FRET imaging on tau biosensors were used to assess the AD-like phenotype related to developmental Pb exposure.

Results: Neural progenitor cells (NPC) were exposed to environmentally relevant levels of Pb at 15 and 50 ppb for 48 h, differentiated into cortical neurons and assessed for maturity and AD-like pathogenesis. Although developmental Pb exposure does not significantly alter the differentiation process, we observed altered neuronal morphology including neurite length and synapse density. Furthermore, mature neurons with prior Pb exposure showed significantly elevated intracellular calcium concentration, as well as an altered epigenetic landscapes and transcriptomic profile. Comparison of transcriptomic data between treated and unexposed cells revealed differentially expressed genes enriched in calcium-homeostasis regulating processes such as GRIN1 and CAMK2B which we further verified using immunofluorescence. Developmental Pb exposure also leads to elevated phosphorylated tau in treated neurons and secreted tau aggregates and Aβ 42 in cell culture medium along with other AD transcriptomic markers.

Conclusion: Collectively, our findings provide an evidence base for Ca dysregulation caused by developmental Pb exposure as a plausible molecular mechanism accounting for increased AD risk in populations with developmental Pb exposure.

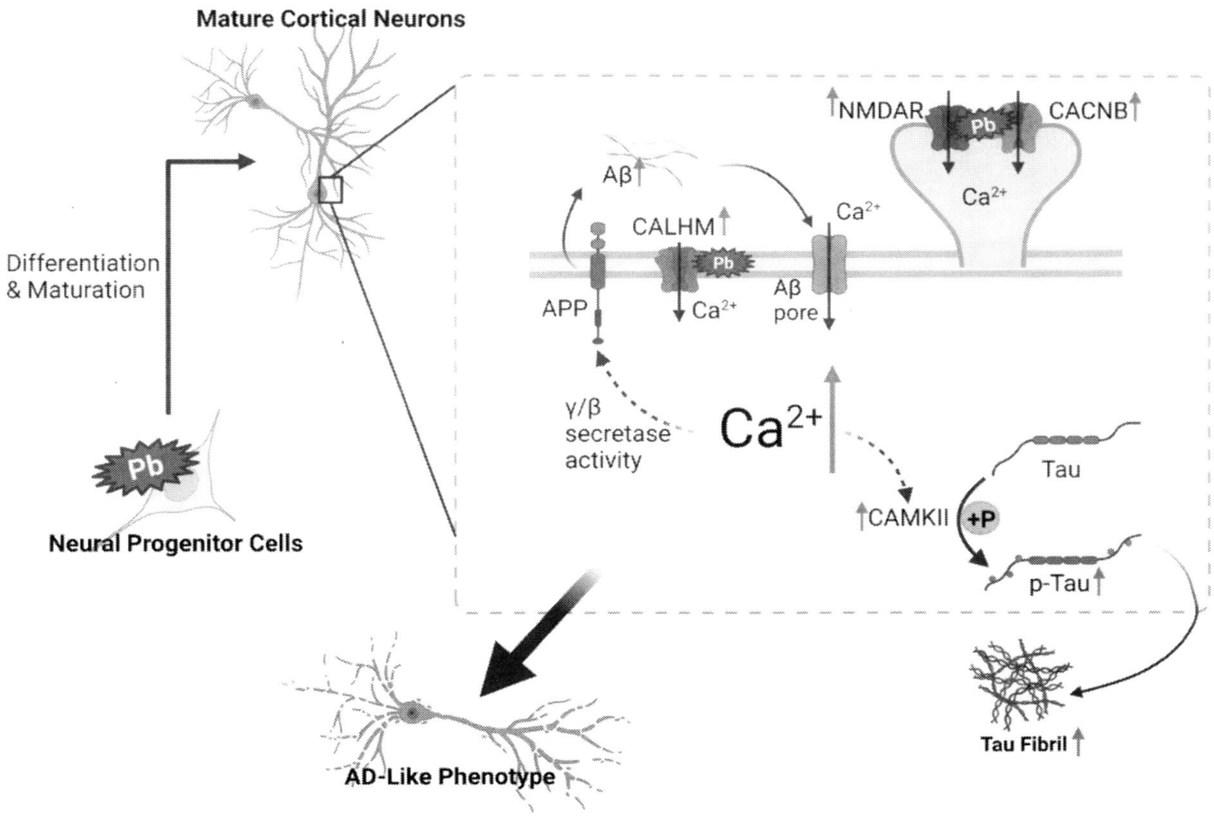
Mature Cortical Neurons
Differentiation & Maturation
Neural Progenitor Cells
Pb
NMDAR
CACNB
Pb
Aβ
Ca2+
CALHM
Ca2+
Pb
APP
Ca2+
Aβ pore
γ/β secretase activity
Ca2+
Tau
CAMKII
+P
p-Tau
AD-Like Phenotype
Tau Fibril

Analysis of Internal Flow Phenomena in a High Speed Rotating Cylinder Using Double Parabolic Axial Flow Model

Sahadev Pradhan

Chemical Technology Division, Bhabha Atomic Research Centre, Mumbai-400 085, India.

ABSTRACT:

In this study we investigate the internal flow phenomena in a high speed rotating cylinder using double parabolic axial flow model for wall pressure in the range 20 to 100 m-bar [1, 2, 3, 4, 5, 6, 7, 8, 9, 10, 11, 12, 13]. These includes the quantitative estimation of the important process parameters L_0, L, m, and (F/L).Here, L_0 is the internal circulation rate at total reflux, L is the actual internal circulation rate, $m = L/ L_0$ is the internal circulation parameter, (F/L) is the internal reflux ratio, and F is the external feed flow rate. The radial profile of axial mass flux (ρv_z) is specified as a function of radial scale height (ξ) by two parabola. One parabola represents the flow downward in the region near the rotor wall, and the second one representing the upflow adjacent to the downflow. The feed gas introduced into the rotating cylinder is considered to be associated with the upflowing stream ((Pradhan & Kumaran, *J. Fluid Mech.*, vol. 686, 2011, pp. 109-159); (Kumaran & Pradhan, *J. Fluid Mech.*, vol. 753, 2014, pp. 307-359)). An important finding is that as the wall pressure is increased from 20 to 100 m-bar, there is a increase in L_0, L, and m. However, there are important differences. L_0 initially increases at a faster rate and then saturates at high wall pressure. On the other hand the parameters L and m increases monotonically with the wall pressure. The effect of feed flow rate on the parameters L_0, L, and m is also studied and the analysis indicates that as the feed flow rate is increased, the parameters L_0, L, and m increases, and the effect is significant at high wall pressure. The analysis also shows that as the average gas temperature is increased, the parameters L_0, L, and m increases, and the effect is more pronounced at high wall pressure. However, with the increase of wall pressure, the internal reflux ratio (F/L) decreases, initially at faster rate and then become constant at high wall pressure, which indicates that there is a strong coupling between the process parameters L_0, L, m, and (F/L), and the results also reveal that at high wall pressure (hence at high holdup) the separation process takes place under improved equilibrium condition.

Keywords: High-speed rotating flow, Double parabolic axial flow, Internal flow phenomena, high wall pressure.

References:

[1] PRADHAN, S. & KUMARAN, V. 2011 The generalized Onsager model for the secondary flow in a high-speed rotating cylinder. *J. Fluid Mech.* 686, 109.

[2] KUMARAN, V & PRADHAN, S. 2014 The generalized Onsager model for a binary gas mixture. *J. Fluid Mech.* 753, 307.

[3] OLANDER, D. R. 1981 The theory of uranium enrichment by the gas centrifuge. *Prog. Nucl.*

[4] WOOD, H. G. & SANDERS, G. 1983 Rotating compressible flows with internal sources and sinks. *J. Fluid Mech.* 127, 299.

[5] WOOD, H. G. & BABARSKY, R. J. 1992 Analysis of a rapidly rotating gas in a pie-shaped cylinder. J. Fluid Mech. 239, 249.

[6] WOOD, H.G. & MORTON, J.B. 1980 Onsager's pancake approximation for the fluid dynamics of a gas centrifuge. *J. Fluid Mech.* 101, 1.

[7] BIRD, G. A. 1994 Molecular gas dynamics and the direct simulation of gas flows. *Clarendon Press*, Oxford.

[8] VINCENTI. 1967 Introduction to Physical Gas Dynamics. Wiley, New York.

[9] BIRD, G. A. 1963 Approach to translational equilibrium in a rigid sphere gas. Physics of fluids 6, 1518.

[10] CERCIGNANI, C. 2000 Rarefied gas dynamics. From basic concepts to actual calculations. Cambridge University Press.

[11] SOUBBARAMAYER. 1979 Centrifugation. In Uranium Enrichment (Ed. S. Villani, Springer) 1,

[12] E. RATZ, One-Stage enrichment with centrifuges, Proceedings of the sixth workshop on gases in strong, Rotation, Tokyo, 621-654 (1985).

[13] E. RATZ, An analytical solution for the separative power of a gas centrifuge, Proceedings of the fifth workshop on gases in strong, Rotation, University of Virginia, Charlottesville (1983).

84ah Soiling Impacts on Hydrophobic Coating with and Without Dew Suppression

Aniket Ratnaparkhi, Drashti Dave, Michael Valerino, Mike Bergin, Chinmay Ghoroi

The deposition of particulate matter (PM) on solar PV surfaces reduces energy generation ranging from 17 to 25 % which results in huge losses to the PV industry. Hydrophobic coatings have been shown to reduce the PM accumulation on PV surfaces. Additionally, covering panels during the night is expected to reduce the PM deposition as well as suppress the dew formation on the surface. The aim of this study is to examine the impact of soiling on hydrophobic coatings with and without dew suppression. In the present study, both uncoated and hydrophobic-coated glass slides were utilized to collect particulate matter (PM) deposition over a 70-day period in a field study. To prevent dew formation, certain uncoated and coated glass slides were covered from sunset to sunrise whereas, some other samples were kept open during the entire sampling cycle to allow dew formation. The findings of the study indicate that the hydrophobic-coated glass slides subjected to dew suppression experienced a higher percentage reduction in soiling (50 ± 13%) compared to those without dew suppression (31 ± 15%). The difference was statistically significant, as indicated by a two-tailed p-value of less than 0.05. The findings of this study also include particle size analysis using optical microscopy, which indicates that the hydrophobic coating under dew suppression can significantly reduce particle deposition (by 60.0 ± 8.1%, mainly in the 10 to 30 μm range) compared to the uncoated slides under dew suppression throughout the entire sampling period. In summary, this study contributes to the current knowledge of soiling mitigation methods for the solar PV industry by exploring the effects of dew suppression on hydrophobic coating. By providing a deeper understanding of this impact, this research could be instrumental in developing more effective strategies for reducing soiling on solar PVs.

84ai Effect of Water-Gas Seepage and Salt Ions on Hydrate Phase Transition in Porous Media

Huiru Sun

Natural gas hydrate is the one of the most potential energy sources to meet the future human development. During the hydrate exploitation process, the multiphase seepage and infiltration of overlying seawater will influent the gas production efficiency. Therefore, it is necessary to reveal the interaction mechanism between multiphase seepage and hydrate phase transition characteristics. In this study, the multiphase flow in hydrate reservoirs is simulated by controlling the seawater–gas flowrate, and the variations of pressure, temperature, and seawater production volume are monitored to during the seawater–gas flow process. The experimental results point out that the hydrate re-formation and accumulation will occur and cause the blockage of flow path. Compared to deionized water-gas flow process, the amount of hydrate re-formation will significantly reduce in the seawater-gas flow system, especially in case of the higher seawater-gas flowrate ratio. Moreover, hydrate re-formation induces the temperature and pressure following a three-stage trend in the continuous seawater-gas flow process: (I) stable stage without hydrate re-formation; (II) rapid increase stage due to mass hydrate reformation; (III) blockage stage where the temperature and pressure obviously decreased and finally kept stable. After blockage, water and gas were stored in pore to continuously re-form solid hydrate barrier to prevent gas overflow, finally a stability reservoirs pressure was obtained.

Effect of water-gas seepage and salt ions on hydrate phase transition in porous media

Huiru Sun*, Mingjun Yang, Yongchen Song

[1] *Key Laboratory of Ocean Energy Utilization and Energy Conservation of Ministry of Education, Dalian University of Technology, Dalian, 116024, PR China*
[2] *State Key Laboratory of Natural Gas Hydrates, Beijing 100027, China*

*Corresponding author(s): H. Sun (huirusun@163.com)

Abstract

Natural gas hydrate is the one of the most potential energy sources to meet the future human development. During the hydrate exploitation process, the multiphase seepage and infiltration of overlying seawater will influent the gas production efficiency. Therefore, it is necessary to reveal the interaction mechanism between multiphase seepage and hydrate phase transition characteristics. In this study, the multiphase flow in hydrate reservoirs is simulated by controlling the seawater–gas flowrate, and the variations of pressure, temperature, and seawater production volume are monitored to during the seawater–gas flow process. The experimental results point out that the hydrate re-formation and accumulation will occur and cause the blockage of flow path. Compared to deionized water-gas flow process, the amount of hydrate re-formation will significantly reduce in the seawater-gas flow system, especially in case of the higher seawater-gas flowrate ratio. Moreover, hydrate re-formation induces the temperature and pressure following a three-stage trend in the continuous seawater-gas flow process: (I) stable stage without hydrate re-formation; (II) rapid increase stage due to mass hydrate reformation; (III) blockage stage where the temperature and pressure obviously decreased and finally kept stable. After blockage, water and gas were stored in pore to continuously re-form solid hydrate barrier to prevent gas overflow, finally a stability reservoirs pressure was obtained.

Keywords: Natural gas hydrate, salt ions, water-gas flow, hydrate re-formation.

84aj Persistence of *Phi6*, a Sars-Cov-2 Surrogate, in Simulated Indoor Environments: Effects of Humidity and Material Moisture Adsorption

Eloise Parry-Nweye, Zhenlei Liu, Yousr Dhaouadi, Xin Guo, Jianshun Zhang, Dacheng Ren

The COVID-19 pandemic brought major challenges to public health and is primarily transmitted via aerosols, droplets, and fomites. However, there is a lack of understanding of virus persistence on fomites (i.e., building materials) under different environmental conditions in a realistic environment. Thus, we aimed to bridge this knowledge gap especially as fomites are a significant virus transmission route. This study utilized *Pseudomonas Phage Phi6*, an 85 nm enveloped dsRNA bacteria virus, as a coronavirus surrogate due to its structural similarities. The experiments were conducted in a 29 m^3 full-scale aerobiology chamber to simulate a realistic indoor environment. The viral titers were quantified via virological Plaque Forming Assays (PFU) and molecular Reverse Transcription Quantitative Polymerase Chain Reaction (RT-qPCR), which quantify the infective viral load and total viral load respectively. We initially assessed the effect of our sampling methods on the viability and integrity of *Phi6* to validate the observed relative humidity (RH)-induced inactivation. Our results show that the infectivity of *Phi6* decreases by $\geq$ 1-log as the RH increases from 25% to 45% to 75% and then increases by $\geq$ 1-log as the RH increases to 85% both in air and when *Phi6* is deposited on surfaces by gravitational settlement. The kinetics of RH-induced inactivation shows the decay rate of infectivity changed in the order (high to low) at 75% » 25% ≅ 45% ≅ 85%. In addition, we show that although material properties may impact virus persistence, local humidity at the air-substrate interface more significantly influences virus persistence on tested surfaces. These results provide more insights into viral transmission under varied environmental conditions with changes in the aerosol pH and how these changes affect the aerosolized virus. These findings will help the design of more effective strategies for viral control in indoor environments.

84ak Multi-Objective Optimization for Work-Integrated Heat Exchange Network

Yongjian Huang, Yu Zhuang, Linlin Liu, Jian Du, Shengqiang Shen

Heat exchange networks and work-heat exchange networks have been effective manners for energy integration. However, the utility system is only considered at the end of streams in traditional work-heat exchange networks, which requires high-level steam and limits the possibility of using multiple utilities. To address the issue, aiming at the coupling of compression, heat exchange network and utility system, this paper proposes an extended superstructure considering inter-stage multiple utilities configuration optimization and both compression and heat exchange on branch streams. A mixed integer nonlinear programming model is established based on the superstructure, which performs multi-objective optimization with the objectives of minimum exergy consumption and minimum total annualized cost. The optimal Pareto solution is obtained by the ε-constraint method. An example with different cases is studied to illustrate the feasibility and efficacy of the proposed model, where the preferable network configurations to weigh the thermodynamic and economic performances are obtained.

84al Pre-Differentiation Pfas Exposure and Its Effect in Human Dopaminergic-like Neurons

Shichen Wu, Junkai Xie, Han Zhao, Xihui Zhao, Chongli Yuan

GenX, also known as hexafluoropropylene oxide dimer acid (HFPO), is a widely used alternative of PFOA starting from 2005. PFOA and GenX belongs to the family of Per- and Polyfluorinated Substances (PFAS), a group of chemicals widely used for water repellent products, non-sticky cookware, and firefighting foams. GenX can now be detected in various water supplies, including surface and ground water as well as air emission, which is raising increasing safety concerns. In 2022, the US Environmental Protection Agency (EPA) issued a health advisory on GenX with a recommended level of 0.01 parts per billion (ppb) in drinking water. Increasing studies have associated GenX with the development of cancers and neurodegenerative diseases; however, few of them have systematically assessed low-dose exposure of GenX and its possible mechanism to disease onset. Utilizing SH-SY5Y, a human cell line that can be differentiated into dopaminergic (DA)-like neurons and have been widely adopted to examine neurotoxic effects, we investigated the effect of GenX exposure prior to differentiation and assessed changes in epigenome, neuronal characteristics, and neurodegeneration markers in DA like-neurons. After prior exposure of low-dose GenX, we perceived nuclear morphology change and chromatin rearrangement in DA-like neurons both before and after differentiation. Alteration in neuron-specific features including changes in neuronal network complexity, decrease in TH level and increase of intracellular Ca^{2+} were observed in mature DA-like neurons after exposure. We also observed an increase in α-Synuclein level and pTau181 level in neurites of GenX-exposed DA-like neurons, which are common markers of neurodegenerative diseases Parkinson's Disease (PD) and Alzheimer's Disease (AD). These results altogether revealed persistent deficits induced by low-dose GenX exposure in DA-like neurons and suggested potential neurotoxicity of GenX that can contribute to disease onset.

84am Sustainable Detection of Oil at Well Site: High Contrast UV Fluorescence Imaging System for Pixel-Level Detection of Oil-Bearing Rock Cuttings

Richa Sharma, Karim Bondabou, Mahdi Ammar, Matthias Francois

In this work, we introduce an economical ultraviolet (UV) fluorescence imaging system for accurate detection of crude oil in rock cuttings. Our basic idea is grounded in the past fluorescence applications in petroleum applications (e.g., prospecting and exploration). Visible fluorescence is initiated by electron energy absorption when excited by UV radiation. Photons of UV radiation collide with the electrons promoting electron excitation from the ground level to a higher energy level (excited state). Subsequently, the electronic energy is released through fluorescence during deactivation back to the ground state. Unsaturated organic structures which contain pi (π) electrons such as aromatics and conjugated polyenes are primarily responsible for the crude oil fluorescence.

Crude petroleum oils are complex mixtures of organic compounds, including saturates, aromatic, polar, and asphaltene. Most downhole spectroscopic fluid measurements are costly and tedious. They are particularly challenging due to the simultaneous presence of varying amounts of gases (CO2, CO, O2, N2, and H2 CH4, etc.). Our work is motivated by the long-standing need of acquiring surface measurements at the well-site as part of the mudlogging activites. This capability can bring huge gains by enabling automation and optimization of the complete workflow based on rapid measurements on each single drilled cutting.

Fluid analysis as part of mud logging workflow can be vital for the first indication of the potential success. Common mudlogger spectroscopy tools are based on Infrared (IR) measurements. However, IR is limited by challenges of overlapping of absorption bands and saturation of the signal due to the high absorptivity of crude petroleum. On the other hand, efforts on advancing the fluorescence-based methods have considered two performance criteria: high sensitivity and faster measurements. Unfortunately, the UV fluorescence inspection that has been used for the past 60 years (for detection of crude oil during mud-logging) does not score well on either criterion. The legacy detection system, called Fluoroscope, provides an eyepiece window to view the interaction of high-power UV radiation and drill cuttings, but it suffers from low robustness and bulky apparatus. The fluoroscope typically provides weak contrast images which are of limited use for reliably applying modern machine learning tools for detection of oil in drilled cuttings. Besides, UV radiation and mercury-based bulbs pose high health and safety risks for the operator. Hence, our claim is that the fluoroscope system is not suitable for integration into automated workflow at well sites. Our proposed UV imaging system targets these bottlenecks and breaks the fixedness of traditional technology.

Our device systematically addresses abovementioned pain points of the conventional oil-show equipment Fluoroscope by improving the safety of user and environment and enabling high-quality data acquisition and integration of measurements for rocks and fluids. More broadly, our UV fluorescence imaging system crucially supports a highly sophisticated quantitative imaging toolbox. The main features enabled by our system include: 1) single pixel-level quantification of oil bearing cuttings; 2) correlative imaging based on merging the white light and UV images for estimation of percent oil bearing cuttings; and 3) Machine learning for rapid oil detection.

84ao A Novel Approach to Develop Industrial Facility's Life-Time Energy Efficient Design Zeeshan *Farooq and Abdulrahman Hazazi, Energy System Division, Pandcsd Saudi Aramco, Dhahran*

Zeeshan Farooq

Industrial facilities are rapidly changing over the years, technical progressions and stratagems have led to break-through its performance. The revolution is brought from new inventions, development of urbane processes/technologies, automation, digitalization, the list is unremittingly growing. Numerous technological advances hard-pressed industries to realize it on the life-cycle basis and reconnoiter sustainability concepts within the scope. On the other hand, the mission to diminish Green House Gas (GHG) emissions to shield environment is required than ever before. Subsequently, the revolution and evolution of products, processes, technologies, and resources, and sustainability obligation to mitigate climatic anomaly congregates on the industrial sector and ordained it to cater an approach to optimize all efforts. The challenge is to develop a methodology by accommodating all the technological advances and sustainability requirements and transform facilities into a better performing asset with reduced emissions. The solution is to make industrial facilities energy efficient which enable them to be competitive and environmentally friendly.

Industrial sector is accountable for approximately one-third of world-wide energy usage that relies on fossil fuels as the primary source of energy. A significant percentage of energy consumed is disbursed to supply utilities for oil & gas facilities so that energy could be engendered for other sectors. The adverse impact of fossil-fuel combustion has instigated efforts to minimize carbon emissions, whilst environmental regulations have a substantial effect on the energy's production cost. Furthermore, it is a competitive world with ever-increasing energy prices, industrial communities are convened to brainstorm over conservation of energy resources and profitability. Energy efficiency should be a foundational strategy to support plans to meet the climate policy goals as well as for being lucrative. Energy efficiency with its integral role in energy intensive industrial processes is critically essential to achieve desired energy-saving potential and reduction in Carbon Dioxide (CO_2) emissions.

Pinch technology imparts a systematic technique for saving energy in processes and on entire sites. This technique helps optimize thermal heat recovery, and immediately became popular as a theoretically elegant yet feasible approach to design Heat Exchanger Networks (HENs). The paper illustrates, through case studies, the 'plant lifetime retrofit-ability concept in industrial plants. It demonstrates the impact of energy demand reduction by identifying design improvements which complement future enhancements, which was coined as "retrofit with retrofit in mind approach". The case study objective is to develop the most energy efficient process design via arranging each future modification in phases for implementation to remain energy efficient throughout its lifecycle without topology change and utilizing minimum capital investments. Each energy saving' modification is implemented in a way that it compliments future identified modifications. Finally, a methodology is proposed to develop inherently sustainable energy efficient design in order to cater the benefits throughout the industrial facility life-cycle.

84ap Economic and Environmental Assessment of Plastic Sorting and Recycling: A Multi-Period Approach

Jaepil Cheon, Junhyeok Son, Yuchan Ahn

Plastic recycling rate reveal low rates compared to production, disposal, incineration rate, largely due to the inadequacy of effective sorting processes [1]. Consequently, plastics are often disposed of through incineration and landfilling, leading to increased greenhouse gas emissions and environmental pollution. Despite several studies that have mathematically approached plastic sorting processes and integrated them with recycling processes for economic and environmental impact comparisons, research applying multi-periods remains scarce.[2]

This study aims to develop an integrated model that considers the growing amount of waste plastic affected by multi-periods and optimizes the benefits of the sorting and recycling processes. A mixed-integer linear programming problem is employed to solve the selection combinations of sorting order and recycling processes for each type of plastic, such as material recycling through pelletization for PET, pyrolysis oil through pyrolysis for LDPE and HDPE, and energy recovery through incineration for other synthetic resins[3]. By combining the sorting process with recycling processes based on sorting order, the study conducts a comparative analysis through several case studies to identify the optimal benefits.

Economic evaluations reveal that when the amount of plastic waste to be processed is minimal, the number of sorting machines has the most significant impact on the objective function. However, as the amount of waste increases, the recycling rate becomes more influential than the number of machines. Environmental evaluations indicate that higher recycling rates result in increased $CO2$ emissions due to greater electricity consumption from the operation of more machines and longer operation times. To address the trade-off between economic and environmental factors, the study selects the optimal model through case studies.

Literature cited:

[1] Geyer, Roland, Jenna R. Jambeck, and Kara Lavender Law. 'Production, use, and fate of all plastics ever made.' *Science advances* 3.7 (2017): e1700782.

[2] Lee, Jaewon, et al. 'Multiobjective Optimization of Plastic Waste Sorting and Recycling Processes Considering Economic Profit and CO2 Emissions Using Nondominated Sorting Genetic Algorithm II.' *ACS Sustainable Chemistry & Engineering* 10.40 (2022): 13325-13334.

[3] Basuhi, R., et al. 'Environmental and economic implications of US postconsumer plastic waste management.' *Resources, Conservation and Recycling* 167 (2021): 105391.

84aq Maximizing Cost Savings and Reducing CO2 Emission in NCC Process with Heat Exchanger Networks: An Integrated Analysis

Subin Jung, Hyojin Jung, Yuchan Ahn

Olefins are considered an essential raw material in the chemical industry to manufacture various products for daily life, including plastic, synthetic fiber, etc. Several techniques have been developed to produce olefins, such as ethane cracking center (ECC), coal to olefin (CTO), and naphtha cracking center (NCC) [1], and NCC is mainly applied in Asia and Europe. In the NCC process, naphtha is thermally cracked at temperature above 800 ◦C and rapidly cooled to temperature below 400 ◦C through a series of step [2]. Since it requires a significant amount of heat and cooling energy simultaneously, the NCC process is energy-intensive and waste energy exists inside of the process. Recycling waste heat and cooling energy is necessary to improve energy independence and economic benefits.

Motivated by these considerations, the heat exchanger network (HEN) using pinch analysis is presented [3]. This methodology involves the use of a heat exchanger to transfer heating and cooling energy between two streams in a process, ensuring that they reach the required condition. Since the installation cost of the heat exchanger depends on the area of the heat exchanger, an economic evaluation based on the area of the heat exchanger should be conducted. Also, as external energy input decrease, not only economic benefits but also environmental factors is affected. Many studies using HEN have proposed an economic evaluation, but environmental impact assessment has not been considered about HEN by calculating the reduced utility amount. Therefore, an integrated evaluation considering the economy and environment should be conducted.

We focus on the development of multiple heat exchange networks using data extracted from an actual NCC processes and optimal heat exchange networks are derived. In addition, an integrated assessment combining economic and environmental impact assessment is conducted based on the amount of carbon dioxide generated from the varied types of energy generation fuel (e.g., coal, natural gas, biogas, waste plastic, etc.) and energy production methods (e.g., solar power generation). Based on this integrated assessment, the optimal HEN configuration will be determined by minimizing the payback period considering installation cost, utility reduction, and CO_2 emission.

Literature cited:

1. Kim, S. and S. Oh, *Impact of US Shale Gas on the Vertical and Horizontal Dynamics of Ethylene Price.* Energies, 2020. **13**(17): p. 4479.
2. Han, S.Y., et al. *Selective formation of light olefins by the cracking of heavy naphtha.* in *Carbon Dioxide Utilization for Global Sustainability: Proceedings of the 7th International Conference on Carbon Dioxide Utilization, Seoul, Korea.* 2004.
3. Linnhoff, B. and E. Hindmarsh, *The pinch design method for heat exchanger networks.* chemical engineering science, 1983. **38**(5): p. 745-763.

84ar Techno-Economic Assessment of Biomass to Valuable Pyrolyzed Hydrochar Via Hydrothermal Carbonization and Pyrolysis.

Cadianne Chambers, Toufiq Reza, Sumit Sharma, Nirupam Aich

Recent studies have shown the pressing need for biochar for adsorbent technologies to mitigate the serious ongoing wastewater pollution. As traditional biochar synthesis yields adsorbent material with average adsorbing favorable properties, recent research has dived into varying advancements to improve the surface characteristics. The coupling of hydrothermal carbonization (HTC) with traditional pyrolysis has gained traction as a means to introduce attractive binding sites for target impurities without the need of involving chemical activators. This venture has provided high impurity uptake compared to existing adsorbent types further solidifying the demand for pyrolyzed hydrochar. However, limited research remains present on the economic feasibility of implementing hydrothermal carbonization and pyrolysis for the industrial production of pyrolyzed hydrochar from biomass. Therefore, this study proposes the plant scale production of pyrolyzed hydrochar aimed to convert 100 tonne/day corn stover biomass waste. Three different cases were investigated initiating varying pyrolysis temperatures (Cases I = 400 °C, Case II = 600 °C, Case 3 = 800 °C) as studies presented direct correlations between adsorbent quality to pyrolysis temperature. The industrial design which implements the two thermochemical conversion techniques were evaluated to calculate the total capital investment (TCI), manufacturing cost, return of investment (ROI) and breakeven selling price of the pyrolyzed hydrochar. Sensitivity analysis on all three cases were performed to distinguish which of the industrial parameters greatly impacted the breakeven selling price of the pyrolyzed hydrochar.

84as Application of a Developed Techno-Economic Analysis Framework to CO2 Electrochemical Reduction Processes

Claudemi Alves Nascimento, Fernando Lima

The climate crisis has demanded many efforts to develop new technologies to reduce emissions and remove CO_2 from the atmosphere. Electrochemical CO_2 Reduction (eCO_2RR) processes to produce value-added chemicals provide a potential alternative for utilizing captured CO_2. These processes have been the focus of recent studies in the literature [1][2][3]. In particular, most of the current modeling and systems studies involving these processes do not consider process integration and up and downstream processing, which may lead to their incomplete techno-economic assessment. Such assessment is crucial to show the competitiveness of emerging eCO_2RR technologies when compared to conventional processes. This study aims to perform a comprehensive Techno-Economic Analysis (TEA) of an integrated CO_2 electrochemical conversion process to Ethylene and Ethanol production modeled using Aspen Plus®.

The integrated process considered in this work consists of two main parts based on literature [1]: reaction using an electrolyzer and separation of the products and CO_2. The modeled electrolyzer consists of a polymer-based electrolyte membrane configuration that operates at room temperature [3]. Also, the gas-phase products and recycled CO_2 are separated using Pressure Swing Adsorption (PSA) units [4], while the liquid products are separated using a distillation column [3]. To perform the TEA, a framework is developed in Python based on the systematic approach previously proposed by our research group [5]. The developed framework in Python standardizes TEA studies for Electrochemical CO_2 reduction processes and can be applied to assess other chemical processes as well. This framework includes standard economic analysis considering capital costs, operating costs, and depreciation, and evaluates profitability aspects such as net present value (NPV), payback time (PBT), and return on investment (ROI). Moreover, the framework contains other features such as sensitivity analysis, assessment of break-even prices, and new economy of learning concepts [5].

The techno-economic analysis results for the integrated CO_2 electrochemical reduction process will be discussed. Specifically, the evaluation of the process feasibility and trade-offs of using this process in terms of potential improvements in cell performance, stability, and electricity price will be addressed.

References

[1] Singh, S., Stechel, E.B., Buttry, D.A., 2017. Transient modeling of electrochemically assisted CO2 capture and release. Journal of Electroanalytical Chemistry 799, 156–166. https://doi.org/10.1016/j.jelechem.2017.05.045

[2] Orella, M.J., Brown, S.M., Leonard, M.E., Román-Leshkov, Y., Brushett, F.R., 2020. A General Technoeconomic Model for Evaluating Emerging Electrolytic Processes. Energy Technol. 8, 1900994. https://doi.org/10.1002/ente.201900994

[3] Ramdin, M., De Mot, B., Morrison, A.R.T., Breugelmans, T., van den Broeke, L.J.P., Trusler, J.P.M., Kortlever, R., de Jong, W., Moultos, O.A., Xiao, P., Webley, P.A., Vlugt, T.J.H., 2021. Electroreduction of CO2 /CO to C2 Products: Process Modeling, Downstream Separation, System

Integration, and Economic Analysis. Ind. Eng. Chem. Res. 60, 17862–17880. https://doi.org/10.1021/acs.iecr.1c03592

[4] Wiheeb, A.D., Helwani, Z., Kim, J., Othman, M.R., 2016. Pressure Swing Adsorption Technologies for Carbon Dioxide Capture. Separation & Purification Reviews 45, 108–121. https://doi.org/10.1080/15422119.2015.1047958

[5] Gazzaneo, V., Watson, M., Ramsayer, C.B., Kilwein, Z.A., Alves, V., Lima, F.V., 2022. A techno-economic analysis framework for intensified modular systems. J. Adv. Manuf. & Process 4. https://doi.org/10.1002/amp2.10115

84at Asset Intensification: Application of Modelling Tools and Methodologies from a Cdmo Perspective

Filipe Ataide, Jose Luis Santos, Filipe Gaspar

Effective asset utilization is crucial for a Contract Development and Manufacturing Organization (CDMO) to thrive in the pharmaceutical industry. Proper utilization of equipment leads to process and asset intensification, resulting in a range of benefits for both the CDMO and its clients.

From a client's perspective, good asset utilization leads to improved product quality, increased reliability, and reduced costs. This is critical for the success of their pharmaceutical products, as consistent, high-quality products are necessary to meet regulatory requirements and customer . For the CDMO, effective asset utilization leads to increased productivity, reduced maintenance costs, and improved resource utilization that releases manufacturing time that can be used to accommodate new business. This translates into a more efficient, flexible and profitable company, as well as a stronger reputation in the industry.

To achieve asset intensification, CDMOs can employ various methodologies, depending on the level of abstraction of the problem which is being aimed for. For example, in a manufacturing process, can be used to optimize the process parameters to achieve a desired product quality and can also be used to predict the behavior of the equipment under different operating conditions at different scales. On the other hand, if one would like to optimize the asset's availability for production, one area to focus on is the changeover (or change-of-line, COL) planning and how to best address the critical resources (e.g., team members and cleaning stations availability) using a modeling framework (e.g. system dynamics or agent based modeling) in order to minimize the COL duration. This last approach can target different levels of abstraction such as a single equipment train , multiple teams and shared cleaning stations.

In this presentation, we will present models and methodologies that are used to cover different levels of abstraction:

- Shop floor level: optimization of the change-of-line of a single asset line with respect to team member allocation and number of cleaning stations
- Building level: optimization of multiple assets with different number of batches and cleaning schedules, shared team members and cleaning stations
- Company level: Design of future capacity taking into consideration projections of next 5 years business needs. The design will encompass not only main assets, but also auxiliary equipment and human resources needs.

84au Developing Efficient and Sustainable Packaging Processes in the Downstream Operations of the Oil and Gas Sector

Swapana Jerpoth, Barnabas Gao, Robert Hesketh, C. Stewart Slater, Mariano J. Savelski, Kirti Yenkie

Oil and gas is the major industry in the energy market and it involves various upstream (exploration, production), midstream (transporting, storing), and downstream (refining, packaging, selling the finished products) activities (Nelson 1969). Each of these activities involves high risks and therefore significant efforts are made by industries to reduce the impact of their processes on the environment. Companies are exploring systematic operational procedures to manage and improve the greenness of these activities. In this work, we study one such downstream operation from a manufacturing and packaging plant of refined petroleum lubricants. We use a Machine Learning (ML) approach for improving the operational performance and sustainability of their packaging activities.

The performance index of a manufacturing company is evaluated by the company's capability to attain the desired production goals and effectively utilize its available resources. Thus, excellent performance can be achieved by developing and implementing an appropriate manufacturing strategy. The lubricant industry produces over 2000 unique lube oil compositions in a single production year. The manufactured products within a certain limit will have comparable physical properties, similar specifications, and applications. Therefore, these manufacturing plants are categorized as multiproduct plants(Zlokarnik 2001). Multiproduct plants use the same process for the sequential production of different products belonging to the same product family, e.g. engine oil family with products for car engines for different brands.

The multiproduct plants are batch operated to sequentially produce the desired products as per the market demands. Therefore, a special necessity of these plants is to avoid cross-contamination between changeovers and efficiently clean the production vessels, ancillary equipment (valves, fittings, filters) and the associated pipeline system. The straight pipeline sections at the plant are cleaned using polymeric pigs. However, pigging cannot be used for variable diameter pipelines and ancillary equipment having complex design (Stewart 2016). Therefore, a technical solution to this problem is flushing the residual oil in the pipes with new oil that is to be packaged next. This results in the formation of a mixed oil that does not meet the specifications of any of the two batches and is regarded as low value mixed oil. The existing flushing operations lack a standardized procedure which result in generation of large volumes of mixed oil leading to economic losses exceeding millions of dollars annually. Moreover, this leads to poor resource management and with a large energy and environmental footprint. With the growing awareness of sustainability and policies related to pollution prevention and mitigation, it has become a necessity to not only make good quality products but also make their manufacturing and packaging processes environment friendly.

To this end, we present a machine learning approach to strategically optimize the cleaning operations of the multiproduct plants. We considered a case study of the lubricant industry where we have 12 months of data consisting of over 900 features and 300 data points. Out of the total data points, 70% were used for training, 15% for validation, and the remainder for testing. Through a multi regression machine learning algorithm (Ray 2019), we develop an approach for systematically optimizing the flushing times and the associated product volume for a particular changeover. Furthermore, we also analyze the associated environmental impacts. Since the environmental impacts are associated with the mass and energy flows of a product, the respective first principles conservation equations for the system add valuable information. Next, we collect process-related sustainability metrics data via LCA (Life Cycle Analysis) tools such as GREET (Greenhouse Gases, Regulated Emissions, and Energy Use in Technologies) model by Argonne National Lab (Wang 2007), SPIonWeb (Shahzad et al. 2014), and GREENSCOPE (Ruiz-Mercado, Gonzalez, and Smith 2014) to predict the CO_2 emissions, Global Warming Potential and Sustainable Process Index (SPI) (Papadokonstantakis et al. 2016; Klemes et al. 2014; Goedkoop and Spriensma 2001). Through this work, we present an approach to making the operations of the multiproduct plants economically profitable and environment friendly. Figure 1. shows a schematic of our overall approach.

References:

Goedkoop, Mark, andSpriensma. R. 2001. "The Eco-Indicator 99: A Damage Oriented Method for Life Cycle Impact Assessment," January.

Klemes, Jiri J., Petar Sabev Varbanov, Peng Yen Liew, Peng Yen Liew, Petar Sabev Varbanov, and Peng Yen Liew. 2014. *24th European Symposium on Computer Aided Process Engineering: Part a and B*. Oxford, NETHERLANDS, THE: Elsevier. http://ebookcentral.proquest.com/lib/rowan/detail.action?docID=1718627.

Nelson, W. L. 1969. *Petroleum Refinery Engineering*. 4th Edition. McGraw-Hill Book Co.

Papadokonstantakis, Stavros, Paraskevi Karka, Yasunori Kikuchi, and Antonis Kokossis. 2016. "Challenges for Model-Based Life Cycle Inventories and Impact Assessment in Early to Basic Process Design Stages." In , 295–326. https://doi.org/10.1016/B978-0-12-802032-6.00013-X.

Ray, Susmita. 2019. "A Quick Review of Machine Learning Algorithms." In *2019 International Conference on Machine Learning, Big Data, Cloud and Parallel Computing (COMITCon)*, 35–39. https://doi.org/10.1109/COMITCon.2019.8862451.

Ruiz-Mercado, Gerardo J., Michael A. Gonzalez, and Raymond L. Smith. 2014. "Expanding GREENSCOPE beyond the Gate: A Green Chemistry and Life Cycle Perspective." *Clean Technologies and Environmental Policy* 16 (4): 703–17. https://doi.org/10.1007/s10098-012-0533-y.

Shahzad, Khurram, René Kollmann, Stephan Maier, and Michael Narodoslawsky. 2014. "SPIonWEB – Ecological Process Evaluation with the Sustainable Process Index (SPI)." In *Computer Aided Chemical Engineering*, edited by Jiří Jaromír Klemeš, Petar Sabev Varbanov, and Peng Yen Liew, 33:487–92. 24 European Symposium on Computer Aided Process Engineering. Elsevier. https://doi.org/10.1016/B978-0-444-63456-6.50082-X.

Stewart, Maurice. 2016. "Pipeline Pigging." In *Surface Production Operations*, 897–960. Elsevier. https://doi.org/10.1016/B978-1-85617-808-2.00013-4.

Wang, M. 2007. "The Greenhouse Gases, Regulated Emissions and Energy Use in Transportation (GREET) Model."

Zlokarnik, Marko. 2001. *Stirring: Theory and Practice*. Wiley. https://doi.org/10.1002/9783527612703.

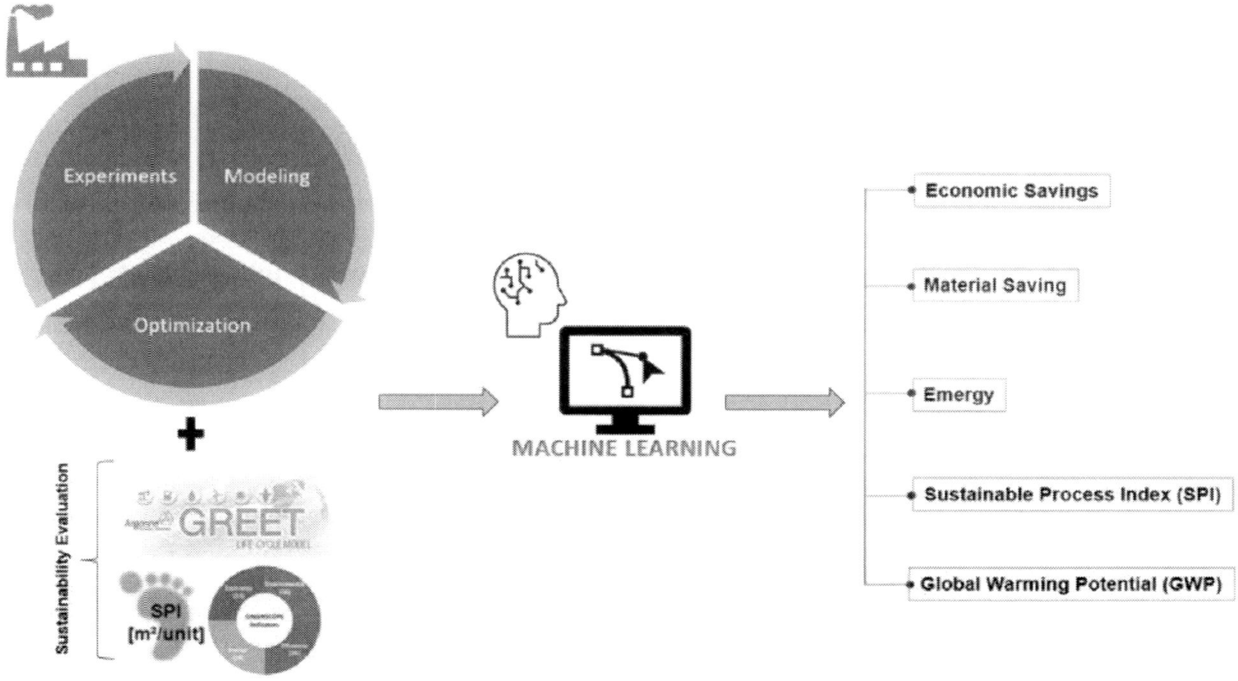

Figure 1: *Schematic diagram showing the ML approach for developing sustainable manufacturing operations of multiproduct plants in the oil and gas sector*

84av Economic Comparison of Nano-Porous Silica Production Processes from Rice Husk and Sand

Semie Kim, Young-Il Lim

Environmental problems caused by global warming continue to make efforts to produce energy from sustainable resources or high value-added products from waste. Rice husk generated during rice milling is an agricultural by-product and contains 18-20 wt% silica. Silica is a compound of silicon and oxygen, and it is used in various applications such as paint ingredients, reinforced rubber additives (tires), concrete mixtures, coating additives, and nano-fibers. In the conventional process, the nano-porous silica is produced from sand, which contains 98% silica. Sodium silicate is produced from silica and sodium carbonate at 1300 °C heated by natural gas. Nano-porous silica (NPS) is produced through several steps in the downstream process. The sodium silicate is combusted at 650-850 °C from rice husk ash, and electricity can be generated exploiting combustion heat.

This study compared the economic feasibility such as total capital investment (TCI) and production cost and the environmental impact for the NPS production processes from sand and rice husks. The NPS production process included combustion, hydrothermal synthesis reactor, ultrafiltration, condensation reaction with H_2SO_4, dryer, wastewater and flue gas treatment. The effect of the NPS production process from rice husk on carbon neutrality was analyzed.

84aw Optimizing Reformer Performance for Integrated Blue Hydrogen-Methanol Production: A Multi-Objective Optimization and Techno-Economic Study

Ajay Koushik V, Achyuta Krishnan, Niket Kaisare, Preeti Aghalayam

Introduction

Utilizing CO_2 as a potential feedstock for producing value-added chemicals is an essential field of research both from environmental and economic perspectives. In this regard, methanol is one of the chemicals that can be produced from CO_2. It has been extensively studied throughout the literature owing to its high demand and wide range of applications. However, producing methanol from CO_2 requires vast quantities of hydrogen (1 mole of CO_2 requires 3 moles of H_2 to make 1 mole of methanol). Several articles in the literature have studied the techno-economic feasibility of conversion of CO_2 to methanol [1,2]; however, most of them considered the source of hydrogen as "green", i.e., hydrogen produced through the hydrolysis of water using renewable electricity. But producing green hydrogen is a technology that is still in a developmental phase. Additionally, if hydrogen is purchased from an external source, ways of storing and transporting it efficiently must be taken care of, further complicating the process. To overcome these challenges, a process scheme is presented in this work where the hydrogen is produced *"in-house"* without the need to purchase/store/transport it.

Integrated Blue Hydrogen-Methanol Synthesis (IBHM)

This process flowsheet contains a methane reformer followed by a methanol reactor. The methane reformer produces syngas from methane, which feeds into the methanol reactor. The hydrogen produced is categorized as blue hydrogen because the CO_2 emissions associated with its production are utilized within the unit. The syngas is mixed with the process CO_2 and sent to the final reactor, where methanol is produced. Although this process has more process equipment than direct methanol synthesis, it eliminates the need to purchase, transport and store external hydrogen.

Reformer Technologies

The IBHM flowsheet has a methane reformer that provides syngas for methanol synthesis. H_2O, CO_2, and O_2 can all be used as oxidants for methane reforming, either individually or in combination, each with its pros and cons. Steam reforming produces syngas of an H_2/CO ratio too high for methanol synthesis. CO_2 is a flue gas that can be 'utilized' for methane reforming (two greenhouse gases react to produce a useful chemical here); however, it has issues such as low-quality syngas and coking on the catalyst surface. Achieving the optimal amount of CO_2 for the reformer and the methanol reactor can be done through optimization. Using oxygen for methane reforming is an exothermic process. It needs to be conducted in a controlled environment, which otherwise would lead to the complete combustion of methane (and some associated safety issues). A combination of two or three of these is being extensively studied in the literature to overcome the shortcomings of these three oxidants individually [3].

Workplan and Expected Outcomes

This work aims to provide a detailed techno-economic and environmental analysis for the optimized IBHM flowsheet. To do so, reformer optimization is necessary as there are several levers, such as the feed flow rates of CH_4, H_2O, CO_2, and O_2 and the reaction operating conditions to control the overall outcomes of the process. In this case, two important objectives are chosen for the optimization of the IBHM flowsheet

1. Maximize CO_2 utilization while decreasing coking on the catalyst
2. Maximize methanol yield while reducing energy requirement

Since this formulates a multi-objective optimization (MOO) problem, there won't be a single solution but rather a collection of 'optimal' solutions from which the best set of solutions (i.e., feed ratios and reactor operating conditions) would be chosen for the techno-economic and environmental impact analysis.

Overall, this work will enable one to understand different reforming processes and their impact on the overall methanol yield. The objectives include several factors of practical relevance, such as coking. This work aims to assist the user in identifying the most appropriate reformer configuration for enhancing methanol throughput while maximizing CO_2 utilization, thereby contributing towards a greener and cleaner future.

References

[1] M. Nizami, N. Slamet, W.W. Purwanto, Solar PV based power-to-methanol via direct CO_2 hydrogenation and H_2O electrolysis: Techno-economic and environmental assessment, Journal of CO_2 Utilization. (2022). https://doi.org/10.1016/j.jcou.2022.102253.

[2] G. Lombardelli, S. Consonni, A. Conversano, M. Mureddu, A. Pettinau, M. Gatti, Process Design and Techno-Economic Assessment of biogenic CO_2 Hydrogenation-to-Methanol with innovative catalyst, Journal of Physics: Conference Series. 2385 (2022). https://doi.org/10.1088/1742-6596/2385/1/012038.

[3] P. Balasubramanian, I. Bajaj, M.M.F. Hasan, Simulation and optimization of reforming reactors for carbon dioxide utilization using both rigorous and reduced models, Journal of CO_2 Utilization. 23 (2018) 80–104. https://doi.org/10.1016/j.jcou.2017.10.014.

84ax Life Cycle Assessment of Biodiesel Production Processes from Waste Animal Fats with Pretreatment Process

Pyeong-Gon Jung, Semie Kim, Young-Il Lim

The demand for alternative energy is increasing due to the recent rise of global warming. Biodiesel is a promising source of renewable energy. Biodiesel production involves several stages, including a transesterification reactor, water washing, drying, and wastewater treatment. This study aims to perform a techno-economic analysis (TEA) and life cycle analysis (LCA) of biodiesel production from animal fat waste, which is an inexpensive feedstock against vegetable oils. However, it contains high concentration of free fatty acid (FFA). Contact between FFA and alkali catalysts causes unwanted saponification reactions of triglycerides and alcohol, resulting in the formation of soap that accelerates catalyst consumption and decreases ester yields. To avert these complications, neutralization reactions were employed.
Based on the process simulation using ASPEN Plus, the economic values such as total capital investment (TCI), total production cost (TPC), return on investment (ROI), and payback period (PBP) were calculated. Furthermore, environmental impacts of the overall process were evaluated using the ISO 14040/14044 standards. The outcomes of this study are expected to provide valuable insights into the application of neutralization reactions in biodiesel production.

84ay Catalytic Performance Promotion of Pd Cluster Towards H_2O_2 Production By Potential-Driven Coordination Adjustment

Zhiping Deng, Xiaolei Wang

Electrochemical synthesis of hydrogen peroxide has been emerging as an appealing process for the onsite production of this chemically valuable oxidant, while its commercial application is still hindered by the development of efficient electrocatalyst. Herein, we design an efficient electrocatalyst for the electrochemical synthesis of H_2O_2 consisting of Pd cluster loading on hollow mesoporous carbon spheres (HMCS). Pd coordination environment is well designed with S in the first coordination sphere and O in the second coordination sphere. The as-prepared material exhibits superior performance with high H_2O_2 selectivity (maximum value: 99%) over a wide potential range of 0 to 0.6 V. Moreover, a high kinetics mass activity (4.059 A mg^{-1} at 0.45V) of Pd_x/HMCS is achieved, which is almost three times larger than the state-of-the-art Pd-Hg alloy catalyst. What's more, in neutral electrolyte, Pd_x/HMCS also a positive onset potential of 0.647 V and high H_2O_2 selectivity of 84-93% within a wide potential range. The excellent performance can be attributed to the unique Pd cluster structure and optimized Pd coordination environment, especially the oxygen modification on HMCS support. In-situ Raman results reveal that under potentials, oxygen atoms transfer from the second coordination sphere (oxygen containing functional groups on HMCS) to the first coordination sphere (Pd-O bond). Density functional theory (DFT) calculations confirm that such oxygen coverage helps optimize the *OOH binding energy on Pd surface, resulting in significantly enhanced catalytic performance towards H_2O_2 production. This work reveals the structure difference between in-situ and ex-situ situations, proposes a new mechanism for the improved performance, and provides new insights into the future catalyst design.

84az Stability and Redox Kinetics of Ti^{4+}/Ti^{3+} for Flow Battery Applications

Maria Bruce, Vijay Ramani

Interest in understanding and improving titanium redox kinetics has surfaced as titanium-based anolytes are explored for energy storage applications. Technologies that harness renewable energy have decreased in cost, creating a demand for viable energy storage technologies that are compatible with intermittent sources[1]. To be practical, the production *and* storage/transmission/distribution of this energy must be less expensive than fossil-based alternatives. Flow batteries can be suitable for long-duration and intermittent storage with the unique benefit of scaling power and energy independently[1,2]. The most developed and mature flow battery is the all-vanadium redox flow battery (RFB). While this technology has undergone commercial production, it is not yet widely available due to high cost, driven by the expense of vanadium itself[1]. There have been several alternative RFB chemistries proposed that utilize titanium-based anolytes paired with various catholytes such as cerium and manganese[2,3]. For the titanium-cerium system, preliminary cycling data shows minimal coulombic efficiency fade but relatively modest energy efficiency (< 70%) at current densities of 150 mA/cm^2 due to kinetic limitations. From comparing the rate constants of the two redox couples (TiO^{2+}/Ti^{3+} and Ce^{3+}/Ce^{4+}) in methanesulfonic acid, the titanium rate constant is three orders of magnitude lower than that of cerium. The addition of Mn^{2+} ions has been explored to improve titanium kinetics in sulfuric acid for a titanium-manganese RFB, but reported no kinetic effects[3]. Both these chemistries are limited by the research gap surrounding titanium redox kinetics. Herein we report experimentally observed titanium redox kinetics with methanesulfonic acid and with sulfuric acid supporting electrolytes. We demonstrate the transient stability of these electrolytes via both electrochemical behavior and spectroscopic shifts. Understanding the kinetics of the titanium redox reaction will pave the way to make further advancements in titanium anolyte RFB systems.

References:

[1] Gü, T. M. (2018). Review of electrical energy storage technologies, materials and systems: challenges and prospects for large-scale grid storage. *2696 | Energy Environ. Sci*, *11*, 2696. https://doi.org/10.1039/c8ee01419a

[2] Shrihari Sankarasubramanian, Yunzhu Zhang, Cheng He et al. An Aqueous, Electrode-Decoupled Redox-Flow Battery for Long Duration Energy Storage, 27 January 2021, PREPRINT (Version 1) available at Research Square [https://doi.org/10.21203/rs.3.rs-150474/v1]

[3] Dong, Y., Kaku, H., Miyawaki, H., Tatsumi, R., Moriuchi, K., & Shigematsu, T. (2017). Titanium-Manganese Electrolyte for Redox Flow Battery. *SEI TECHNICAL REVIEW*, *84*, 35–40.

84ba Highly Efficient Redox Flow Batteries Using Low Cost Materials

Abena Williams, Robert Emmett, Xueting Wang, Eric M. Davis, Mark E. Roberts

Redox flow batteries (RFBs) have gained prominence as energy storage devices in recent years. This is mainly because of their unique design that permits independent scaling of energy and power outputs for grid-scale renewable energy storage. Despite the pronounced advantages RFBs have over other electrochemical energy devices, they are still hindered by material cost relative to performance. The widely commercialized vanadium redox flow battery (VRFB) is restricted by the limited availability and expensive cost of the vanadium transition metal as well as the Nafion^{TM} membrane which accounts for about 40% of the cell stack cost. Thus, to increase the viability of flow batteries for mainstream grid scale renewable energy storage, there is a need to utilize cheaper and more efficient materials. Previously, we investigated the effect of redox-active carbon electrodes on the reaction kinetics at the electrode-electrolyte interface of low cost redox chemistries. Carbon electrodes containing redox-active iron nanoparticles yielded approximately 2.5x higher power densities and 57% higher energy densities compared to inert carbon electrodes in coin cells using an all iron redox chemistry electrolyte. Moreover, we showed that iron functionalized carbon electrodes improve the power density of the low cost and highly electroactive zinc iodine redox flow battery (ZIRFB) by 100% compared to inert carbon electrodes and had 90% lower charge transfer resistances.

In this work, we seek to further develop a high performing and low cost redox flow battery technology by examining the effect of the cation exchange membrane (CEM) on the electrochemical performance of a zinc iodine RFB. Sulfonated Poly (Ether Ether Ketone) SPEEK membranes were prepared by sulfonation of PEEK powder with various degrees of sulfonation (DS). The DS of SPEEK was characterized by proton nuclear magnetic resonance (^{1}H NMR). Additionally, the ion exchange capacity (IEC) and equilibrium water uptake (EWU) of the membranes were measured. Lab scale ZI RFBs were assembled with 1.0 M electrolyte concentration, iron functionalized carbon electrodes, SPEEK membranes and then with Nafion^{TM} 212 as a comparison of the membrane's performance. Battery testing included conditioning at 5.6 mA/cm^2, polarization analysis and cycling at 17 mA/cm^2. SPEEK membranes were observed to have higher equilibrium water uptakes and ion exchange capacities than nafion 212 indicating that SPEEK is ideal for cation transport. In the ZI RFB, SPEEK had a 370% increase (8.0 Wh/L to 1.7 Wh/L) in energy density compared to nafion 212. Additionally, the cycling columbic efficiency was 92.9% with SPEEK compared to 75.9% with nafion 212. From polarization analysis, a 24% increase in peak power density was obtained. This confirms SPEEK's high cationic selectivity and makes it an ideal CEM for the ZI RFB. The results coupled with the economic merits of SPEEK, and the zinc iodine redox chemistry provide a pathway for low cost, highly efficient grid scale renewable energy storage.

84bb Understanding the Role of Calcium Zincate ($CaZn_2(OH)_6 \cdot 2H_2o$) in Improving Cycle Life of Rechargeable Alkaline Zinc Batteries

Patrick K. Yang, Damon E. Turney, Michael Nyce, Timothy N. Lambert, Stephen O'Brien, Sanjoy Banerjee, Gautam Yadav, Jinchao Huang, Meir Weiner, Shinju Yang

There is a growing need for safe, affordable, low cost, and reliable grid scale energy storage as the world aims to cut down on fossil fuels and shift towards renewable sources of energy. Renewables are intermittent and need to be coupled with energy storage devices to allow energy generation to match hours of largest demand. Metallic zinc (Zn) is investigated and produced industrially for anodes in primary and rechargeable Zn batteries such as Zn/Ni, Zn/Air, Ag/Zn, and Zn/MnO_2. Zinc chemistries provide a high theoretical capacity, relative abundance, non-toxic, and non-flammable nature which make it safe for energy storage.

Metallic zinc itself has shown technological difficulties via poor reversibility when used alone as an anode in alkaline electrolytes. Failure mechanisms are passivation, shape change/redistribution, dendrite formation, hydrogen evolution, and the crossover of zincate ($Zn(OH)_4$) into the cathode. Zinc oxidizes that form during discharge include the zincate ion which is dissolved in the electrolyte and also solid zinc oxide (ZnO). Zincate ions that remain in solution, i.e., not lost to side products or irrecoverable ZnO precipitates or cross over to the cathode, are then able to participate in the cycling process back to metallic Zn on charge.

ZnO anodes with various additives have also been previously investigated to improve the cyclability of metallic zinc. The addition of calcium hydroxide ($Ca(OH)_2$) into ZnO anodes has shown great promise to improving the overall cyclability of the anode. It is widely understood that $Ca(OH)_2$ helps boost performance of ZnO anodes due to the in-situ formation of calcium zincate ($CaZn_2(OH)_6 \cdot 2H_2O$) while cycling the cell. It has been shown that calcium zincate yields a lower solubility of zincate in alkaline KOH electrolytes compared to just ZnO. The lower solubility helps to reduce the dissolution of zincate ions which is susceptible to the losses mentioned above.

There has been some work done in the literature regarding anodes composed primarily of pure Calcium Zincate, with varying electrolyte amounts and other additives to produce varying performance. Making a direct comparison of these pure Calcium Zincate anodes with each other or with a commercial zinc anode is difficult to do because of these experimental differences that cause changes in performance. Ex-situ synthesized calcium zincate anodes show promise over metallic zinc and ZnO anodes at high zinc utilization, but the commercial production of Calcium Zincate powders is still relatively small. We would like to understand the optimal amount of Calcium Zincate powders to use as an additive to commercial metallic zinc anodes to minimize additives while improving the cycling performance.

Various anode formulations with different proportions of Zinc, Zinc Oxide, and Calcium Zincate were fabricated with standardized electrode size, theoretical capacity, separators, and electrolyte concentration at high 50% zinc utilization to understand the effects on battery cycling. These Zinc, Zinc Oxide, and Calcium Zincate anodes were cycled vs two sintered nickel cathodes galvanostatically between 1.2 to 1.98 volts and studied charged, discharged, and in between using cycle life, coulombic/energy efficiency, XRD, SEM, and EDAX mapping to understand morphology and chemical changes within the anodes. Deeper understanding of calcium zincate's physical and chemical mechanism can help to explain the improved performance as an anode material and its addition can help to increase the cycle life and higher utilization of zinc in rechargeable secondary zinc batteries.

Acknowledgements:

This work was supported by the U.S. Department of Energy Office of Electricity. Dr. Imre Gyuk, Director of Energy Storage Research at the U.S. Department of Energy Office of Electricity, is thanked for his financial support of this project. This work was performed, in part, at the Center for Integrated Nanotechnologies, an Office of Science User Facility operated for the U.S. Department of Energy (DOE) Office of Science. This work was also supported by the United States Nuclear Regulatory Commission under grant number NRC- HQ-60-17-G-0030. The views expressed in this article do not necessarily represent the views of the U.S. Department of Energy or the United States Government. This article has been co-authored by an employee of National Technology & Engineering Solutions of Sandia, LLC under Contract No. DE-NA0003525 with the U.S. Department of Energy (DOE). The employee owns all right, title and interest in their contribution to the article and is solely responsible for its contents. The United States Government retains and the publisher, by accepting the article for publication, acknowledges that the United States Government retains a non-exclusive, paid-up, irrevocable, world-wide license to publish or reproduce the published form of this article or allow others to do so, for United States Government purposes. The DOE will provide public access to these results of federally sponsored research in accordance with the DOE Public Access Plan https://www.energy.gov/downloads/doe-public-access-plan.

84bc Effects of Composite Sulfur Electrode Structures and Electrolyte Compositions on Rechargeable Aluminum-Sulfur Batteries

Snehal Bhalekar, Robert Messinger

Aluminum-sulfur (Al-S) batteries are an emerging electrochemical energy storage technology with great promise due to the earth abundance, very low cost, high capacity, and safety of the electrode materials. Recently, we revealed by nuclear magnetic resonance (NMR) spectroscopy that soluble electrolyte-coordinated sulfur species form upon electrochemical discharge, resulting in loss of active mass that results in rapid specific capacity fade.[1] Pang et al. also showed that the large overpotential in Al-S batteries can be mitigated by operating the batteries at high temperatures (e.g. 100 °C) and by using a molten salt electrolyte instead of a room-temperature chloroaluminate ionic liquid electrolyte.[2] However, capacity fade associated with the dissolution of aluminum-polysulfides remains a challenge. Here, we study different carbon-sulfur composite electrode structures, characterize electrolyte speciation and ion transport properties, and investigate the effects of electrolyte additives to improve specific capacity retention and the energy efficiency of cycling over a range of operating temperatures. Composite sulfur electrodes were prepared with different sulfur contents, different carbon structures, and different preparation methods to study how they affect the presence of electrolyte-soluble polysulfide intermediates. Rechargeable Al-S batteries were studied using two different electrolytes: a molten salt consisting of a eutectic mixture of $AlCl_3$, NaCl, and KCl, and an ionic liquid consisting of a Lewis acidic mixture of $AlCl_3$ in 1-ethyl-3-methylimidazolium chloride ([EMIm]Cl). The species in the electrolyte were identified and their relative populations were quantified by liquid-state ^{27}Al, ^{1}H, and ^{13}C NMR spectroscopy, while ion dynamics and diffusion were characterized by NMR relaxation and pulsed-field-gradient (PFG)-NMR diffusion measurements. The effects of electrolyte additives were studied on electrolyte properties and electrochemical interfaces. In aggregate, the results yield insights into how composite sulfur electrode compositions and structures, as well as electrolyte compositions and ion transport properties, influence the electrochemical properties of rechargeable Al-S batteries.

References:

1. "Soluble electrolyte-coordinated sulfide species revealed in Al-S Batteries by nuclear magnetic resonance spectroscopy," Rahul Jay, Ankur L. Jadhav, Leo W. Gordon, Robert J. Messinger, *Mater.*, 2022, *34*, 4486-4495.
2. "Fast-Charging Aluminium–Chalcogen Batteries Resistant to Dendritic Shorting.," Quanquan Pang, Jiashen Meng, Saransh Gupta, Xufeng Hong, Chun Yeun Kwok, Ji Zhao, Yingxia Jin, Like Xu, Ozlem Karahan, Ziqi Wang, Spencer Toll, Liqiang Mai, Linda Nazar, Mahalingam Balasubramanian, Badri Narayanan, Donald Sadoway, *Nature.*, 2022, *608*, 704-711.

84be Efficient Scalable Hydrothermal Synthesis of MnO_2 with Controlled Polymorphs and Morphologies for Enhanced Battery Cathodes

Shifeng Hong, Shuo Jin, Lynden A. Archer

MnO_2 is a versatile, cost-effective transition metal oxide that has attracted interest in multiple domains, including as an active cathode material or catalyst for electrochemical energy storage in batteries. Hydrothermal methods are among the most efficient approaches for MnO_2 synthesis. These approaches enable facile, versatile production of MnO_2 in any of its crystalline phases (α, β, δ, etc.), with the dominant product being determined by reaction conditions such as precursor concentration and temperature. These benefits unfortunately come with impractically low product yields (~9%) and long reaction times. Here, we report that low-molecular weight, water-soluble polymers function as effective nucleating agents for the hydrothermal synthesis of MnO_2. For fundamental reasons, these polymers are also reported to promote large (10- to 40-fold) increases in reaction rate and yield. We evaluate the physical and crystallographic features of the synthesized MnO_2 and find that depending upon the reaction conditions, the polymer-assisted synthesis yields dominantly δ-MnO_2 or β-MnO_2. The as-prepared δ-MnO_2 materials with extra interlayer water are studied as electrodes for aqueous Zn-ion battery applications and are shown to support long-duration storage at both moderate and high rates.

84bf Imidazole-Based Concentrated Hydrogen-Bonded Electrolytes for Energy Storage Applications

Miguel Muñoz Sánchez, Burcu Gurkan, William Dean

Recently, Concentrated Hydrogen Bonded Electrolytes (CoHBEs) have emerged as promising alternatives to conventional electrolytes, owing to their wide electrochemical stability windows, high energy density, low volatility and high proton conductivity.[1] Similarly, imidazole-based systems have been recognized as promising CoHBEs because of their intrinsic hydrogen-bonding network, which enhances proton conductivity by supporting Grotthuss proton diffusion. However, their practical application is currently limited by their high viscosity and solid-state behavior at room temperature (RT). In this study, we explored the potential use of imidazole derivatives and redox-active acids to improve proton conduction in CoHBE systems by leveraging the H-bonding network and proton conduction for Proton-Coupled Electron Transfer (PCET) reactions. The addition of acidic species to imidazole-based systems increases the conductivity by promoting the dissociation of the antiparallel supramolecular structure of the imidazole chains, enhancing the charge transfer capacity of the electrolyte, as described by Cosby *et al.*[2,3]. Furthermore, the redox-active nature of the acid species allows for reversible oxidation and reduction reactions, enabling them to act as electron shuttles and promoting enhanced PCET capacity within the electrolyte.

In preliminary studies, hydroquinone sulfonic acid (HQSA) and 3-amino-4-hydroxybenzoic acid (AHBA) were selected as quinone and azaquinone acid derivatives, respectively, based on their reported redox behavior in acidic electrolytes.[4,5] Several molar compositions (8:1, 4:1, 1:1) of the imidazole derivatives and the redox-active species were tested to form liquid mixtures at RT and the electrochemical behavior of the resulting liquid mixtures was evaluated by cyclic voltammetry. Liquid mixtures are more likely to form at RT when the molar concentration of imidazole is significantly greater than that of the acid (8:1). Liquid imidazole-based supporting electrolyte was necessary to improve the solubility of the mixtures required for the electrochemical characterization by cyclic voltammetry (CV). Electrochemical studies revealed that the redox activity of the acidic species was maintained in some samples at RT. However, it is important to consider the variation in the redox potentials ($E_{1/2}$) towards negative values in comparison with the $E_{1/2}$ values reported for the aforementioned aqueous acidic electrolytes.[4,5] The observed negative potential values are attributed to the basic nature of imidazole when mixed with acidic species, where imidazole acts as a proton acceptor, neutralizing the acid and forming a salt. This can change the pH and affect the redox behavior of the components in the mixture. Therefore, when designing and optimizing these mixtures for use as electrolytes, careful consideration of the concentration and selection of acid species is necessary.

References:

1. Ghahremani, R., Savinell, R. F. & Gurkan, B. Perspective—Hydrogen Bonded Concentrated Electrolytes for Redox Flow Batteries: Limitations and Prospects. *J. Electrochem. Soc.* **169**, 030520 (2022).
2. Cosby, T., Holt, A., Griffin, P. J., Wang, Y. & Sangoro, J. Proton Transport in Imidazoles: Unraveling the Role of Supramolecular Structure. *J. Phys. Chem. Lett.* **6**, 3961–3965 (2015).
3. Cosby, T., Vicars, Z., Heres, M. & Sangoro, J. Associating Imidazoles: Elucidating the Correlation between the Static Dielectric Permittivity and Proton Conductivity. *Phys. Rev. Lett.* **120**, 136001 (2018).
4. Beiginejad, H., Nematollahi, D. & Varmaghani, F. Electrochemical Oxidation of Some Aminophenols in Various pHs. *J. Electrochem. Soc.* **160**, H41 (2012).
5. Zhang, Z. J. & Chen, X. Y. Illustrating the effect of electron withdrawing and electron donating groups adherent to p-hydroquinone on supercapacitor performance: The cases of sulfonic acid and methoxyl groups. *Electrochimica Acta* **282**, 563–574 (2018).

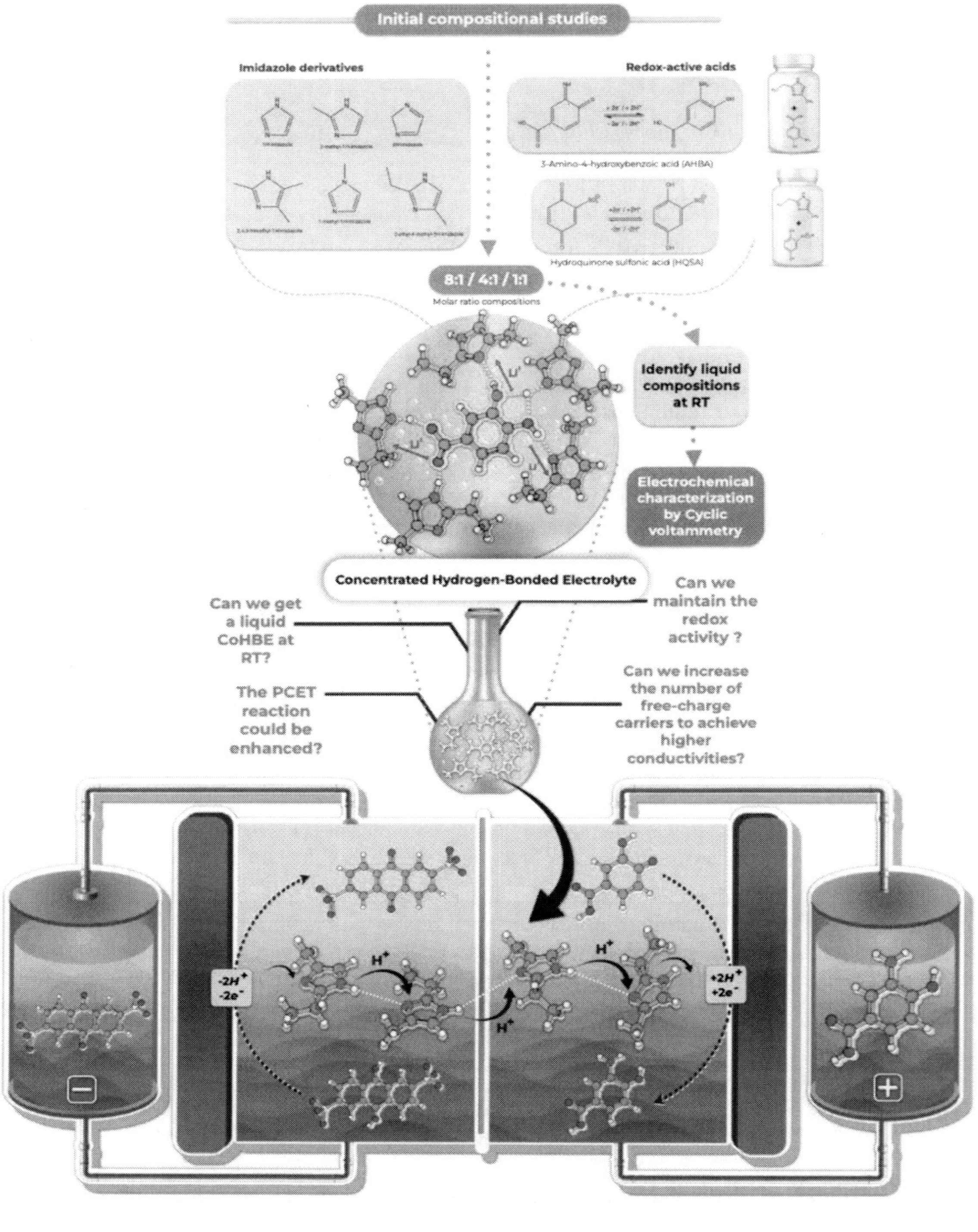

Initial compositional studies
Imidazole derivatives
Redox-active acids
3-Amino-4-hydroxybenzoic acid (AHBA)
Hydroquinone sulfonic acid (HQSA)
8:1 / 4:1 / 1:1
Molar ratio compositions
Identify liquid compositions at RT
Electrochemical characterization by Cyclic voltammetry
Concentrated Hydrogen-Bonded Electrolyte
Can we get a liquid CoHBE at RT?
Can we maintain the redox activity ?
The PCET reaction could be enhanced?
Can we increase the number of free-charge carriers to achieve higher conductivities?
-2H+ -2e-
H+
H+
H+
+2H+ +2e-

84bg Effect of Cathode Precursor Particle Size Distribution on Cobalt-Free Lithium-Nickel-Manganese-Oxide Battery Performance

Sunuk Kim, Sourav Mallick, Arjun Patel, Jethrine Mugumya, Sophie Kothe, M. Parans Paranthaman, Michael L. Rasche, Ram Gupta, Herman Lopez, Mo Jiang

With increasing demands of lithium-ion batteries for electric vehicles (EVs), there is growing need to reduce environment waste and cost[1–3]. Cobalt-free lithium-nickel-manganese-oxide is one eco-friendly cathode material. It is still not well understood whether and how size distribution of precursor microparticles for cathode material affects lithium-ion battery performance. This study first generate oxalate precursor microparticles of uniform and tunable sizes based on slug flow reactor/crystallizer designs[4–8]; then combine these precursor microparticles towards designed size distribution for characterization, such as specific capacity and cycling performance.

References

(1) Darbar, D.; Malkowski, T.; Self, E. C.; Bhattacharya, I.; Reddy, M. V. V.; Nanda, J. An Overview of Cobalt-Free, Nickel-Containing Cathodes for Li-Ion Batteries. *Mater. Today Energy* **2022**, *30*, 101173.

(2) Deng, H.; Belharouak, I.; Cook, R. E.; Wu, H.; Sun, Y.-K.; Amine, K. Nanostructured Lithium Nickel Manganese Oxides for Lithium-Ion Batteries. *J. Electrochem. Soc.* **2010**, *157* (4), A447.

(3) Mallick, S.; Patel, A.; Sun, X.-G.; Paranthaman, M. P.; Mou, M.; Mugumya, J. H.; Jiang, M.; Rasche, M. L.; Lopez, H.; Gupta, R. B. Low-Cobalt Active Cathode Materials for High-Performance Lithium-Ion Batteries: Synthesis and Performance Enhancement Methods. *J. Mater. Chem. A* **2023**, *11* (8), 3789–3821.

(4) Jiang, M.; Zhu, Z.; Jimenez, E.; Papageorgiou, C. D.; Waetzig, J.; Hardy, A.; Langston, M.; Braatz, R. D. Continuous-Flow Tubular Crystallization in Slugs Spontaneously Induced by Hydrodynamics. *Cryst. Growth Des.* **2014**, *14* (2), 851–860.

(5) Nightingale, A. M.; Phillips, T. W.; Bannock, J. H.; de Mello, J. C. Controlled Multistep Synthesis in a Three-Phase Droplet Reactor. *Nat. Commun.* **2014**, *5* (1), 3777.

(6) Mou, M.; Jiang, M. Fast Continuous Non-Seeded Cooling Crystallization of Glycine in Slug Flow: Pure α-Form Crystals with Narrow Size Distribution. *J. Pharm. Innov.* **2020**, *15* (2), 281–294.

(7) Mou, M.; Patel, A.; Mallick, S.; Jayanthi, K.; Sun, X. G.; Paranthaman, M. P.; Kothe, S.; Baral, E.; Saleh, S.; Mugumya, J. H.; et al. Slug Flow Coprecipitation Synthesis of Uniformly-Sized Oxalate Precursor Microparticles for Improved Reproducibility and Tap Density of Li(Ni0.8Co0.1Mn0.1)O2 Cathode Materials. *ACS Appl. Energy Mater.* **2022**, 0–11.

(8) Mou, M.; Patel, A.; Mallick, S.; Jayanthi, K.; Sun, X.-G.; Paranthaman, P.; Kothe, S.; Baral, E.; Saleh, S.; Mugumya, J. H.; et al. Slug Flow Coprecipitation Synthesis of Uniformly-Sized Oxalate Precursor Microparticles for Improved Reproducibility and Tap Density of Li(Ni 0.8 Co 0.1 Mn 0.1)O 2 Cathode Materials. *ACS Appl. Energy Mater.* **2023**, *6*, 3213–3224.

84bh High-Throughput Screening of Electrosynthetic Reactions Enabled By Wireless Electrochemical Cell

Yiming Mo

Electrochemical synthesis has recently emerged as an environmentally benign method for synthesizing value-added fine chemicals. Its unique reactivity has attracted significant interests of synthetic chemists to develop new redox chemistries. However, compared to conventional chemistry, the increased complexity caused by electrode materials, supporting electrolytes, and setup configurations create obstacles for efficient reaction discovery and optimization. One major challenge hindering the development of high-throughput screening (HTS) techniques is that electrochemistry requires electrical connections into the electrolyte to allow external power supply to drive the reaction. This requirement significantly complicates the design of HTS devices for electrosynthesis and limits the corresponding throughput of the system. In this talk, we will present a concept of utilizing wireless power transfer technology to drive electrochemical reactions without direct wire connections. This is enabled by the inductive coupling mechanism, allowing electricity to transfer between coils wirelessly. The coil inside the reacting electrolyte is protected by a miniaturized chemical resistant chassis, and electrodes of different materials can be securely connected with the coil. This conceptual design can significantly simplify the preparation procedure to start an electrochemical reaction. We demonstrated this design with several recently developed electrosynthetic reactions. In addition, we built a fully automated robotic system with 4-channel wireless electrochemical cells to show that this design is HTS-friendly.

84bi Increasing Data Collection Efficiency through Incorporation of Derivative and Uncertainty Information into Gaussian Process Regression

Jacob Monroe, William P. Krekelberg, Austin McDannald, Vincent K. Shen

Molecular simulations generate vast amounts of information, including molecular-level details that, when appropriately averaged, provide estimates of structural, thermodynamic, and transport properties of materials and fluids. Determining the behavior of these properties as a function of state conditions or other adjustable simulation parameters, such as those related to the force field, requires many simulations across state/parameter space. While this is typically a costly endeavor, Gaussian Process Regression (GPR) techniques have recently shown promise as robust models for predicting property behavior over a given state/parameter space and efficiently directing new data collection based on active learning. We show that this automated data collection process can benefit significantly from standard uncertainty estimates, as well as derivative information. It is best practice to include the former in any simulation results, while the latter is often-neglected but can be straight-forwardly obtain from statistical mechanical relations. As an example, we highlight the collection of equation of state data, where relationships between derivative information and property fluctuations are familiar. In this case, both simulations and experiments produce derivative information that can be beneficially incorporated into GPR models. However, we reveal that it is important to assign uncertainty estimates to this information, as certainty is expected to change drastically and systematically with derivative order (e.g., uncertainties in heat capacities may be much higher than those in average energies given a fixed amount of simulation time). We provide an extensible code-base incorporating derivative information and uncertainties into GPR models and demonstrate how this can be used to more efficiently collect both simulation and experimental data. Finally, we will demonstrate how active learning routines can be viewed as improved versions of common data collection algorithms, presenting, as an example, an adaptive, uncertainty-controlled version of Gibbs-Duhem integration.

84bj Data-Driven Design of Selective Partial Agonist for Cannabinoid Receptors

Soumajit Dutta, Austin Weigle, Diwakar Shukla

Drug discovery is a time-consuming and expensive process consisting of a series of steps from target identification to the final approval. The molecule space of potential drug candidates is estimated to be more than 10^{60}. However, screening the entire space with existing computational techniques (e.g., docking) is impossible. Moreover, existing techniques suffer from low hit-rate due to a lack of force-field accuracy. In this work, using deep learning based data-driven modeling, our objective is to design selective partial agonists for cannabinoid receptor 1. These designed molecules will be able to tackle the selectivity and overstimulation issues of existing cannabinoid ligands. The selectivity of the selective partial agonists is decided based on the difference of the predicted binding affinity (pK_i) of cannabinoid receptors 1 and 2 (CB1 and CB2). The functionality of these molecules is determined based role of the molecules in activating the receptor (agonist/antagonist). Furthermore, partial agonism is decided based on the downstream efficacy of the molecules (E_{max}).

We create a predictive data-driven workflow for selectivity, functionality and partial agonism using the existing database of cannabinoid assays for binding and function. The latent space obtained from the junction tree variational autoencoder (JTVAE) is used to featurize the molecules for predicted models. This VAE is pretrained on all class A GPCR ligands (~160K molecules). Based on the stochasticity of the latent space, a data augmentation strategy is used for training of prediction models. This strategy is shown to improve the accuracy of the prediction results significantly. The random forest algorithm is used with ten-fold cross-validations for classification and regression. The overall performance (R^2score) of our regression models is more than 0.96, and the accuracy of the classification models is above 99%.

Selective partial agonists are defined as molecules with more than a 100-fold affinity difference between CB1 and CB2 and being agonists with an Emax value of less than 70%. We screen around one billion compounds in the ZINC20 and ChEMBL databases using our strategy. Around 250 molecules are obtained from the screening process as selective partial agonists for cannabinoid receptors. Further analysis with interpretability shows features crucial for distinguishing partial agonists from unselective molecules. This assists us in generating molecules from pre-trained JTVAE moving along the critical features in the VAE latent space. The predicted and generated selective molecules will be tested experimentally by our collaborator. Ultimately, this research will lead to the discovery of new scaffolds for developing selective partial agonists for therapeutic drug development.

84bk Removal of Insecticides from Wastewater Using Ionic Liquids: A Computational Study

Mustafa Nasser, Mohammed Al Hassan, Ibnelwaleed Hussein, Muneer Ba-Abbad, Imran Khan

Insecticides play a major role in contaminating water sources and making them unfit for use. They are considered toxic and hazardous, and can enter the food chain. To overcome this issue, ionic liquids with their distinguished thermophysical properties are considered green and efficient solvents. In this study, a number of 165 ILs of different combinations of 33 cations and 5 anions were screened using the simulation software the COSMO-RS (Conductor-like Screening Model for Real Solvents) to extract organochlorine insecticides from water by predicting the selectivity and the capacity of ILs at infinite dilution. Density functional theory (DFT) was implemented with COSMO-RS to study the chemical nature on insecticides. Benzene hexachloride (BHC), Heptachlor, Aldrin, Gamma-Chlordane (γ-Chlordane), Endrin, and Methoxychlor are the chosen organochlorine insecticides compounds to be studied. Results show that Endrin and Methoxychlor reflect their chemical nature to be strong H-bond acceptors and weak H-bond donors, while other compounds act as H-bond donors with no possibility of being H-bond acceptors. Besides, it is found that ILs consisting of halides and heteroatomic anions combined with cations provide better selectivity and capacity for insecticides. However, the increase in the alkyl chain length of some ILs based-imidazolium decreases the selectivity and capacity as shown in Figure 1. Moreover, the hydrophobic phosphonium-based ILs deliver better selectivity and capacity to insecticides. In addition, ILs are considered green and efficient solvents to extract insecticides from water.

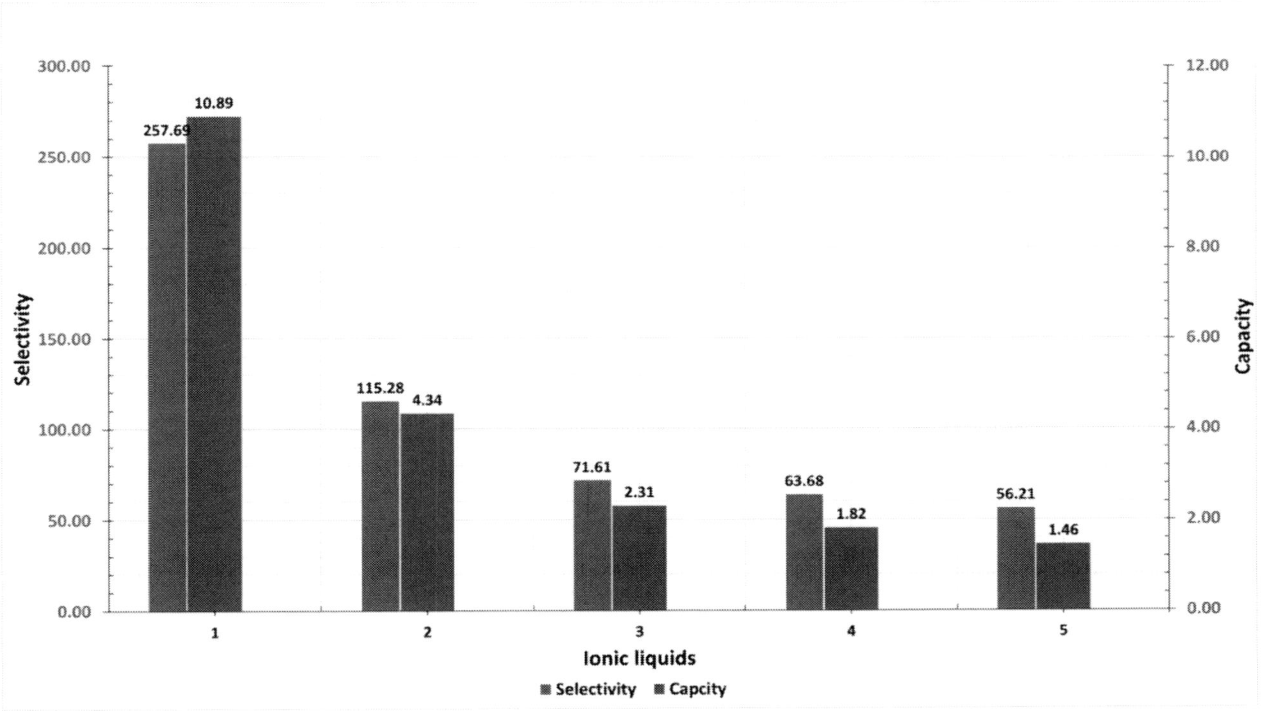

84bl Understanding Proton-Coupled Electron Transfer on Polyoxovanadate Nanoclusters

Andreas Towarnicky, Giannis Mpourmpakis

Poly-Oxo-Metalates (POMs) comprise a versatile and fruitful material group, possessing tunable properties such as H–bond strength/reactivity (e.g. via heteroatom doping)[1], dynamic Proton-Coupled-Electron-Transfer (PCET), and having many applications in the field of small-molecule activation[2]. Great promise for POMs abounds in a wide variety of applications. For example, recent electrochemical research has found interesting PCET behavior and selectivities for the Oxygen Reduction Reaction (ORR) on certain vanadium oxide nanoclusters[3,4].

In this work, we report on the investigation of the experimentally synthesized V_6O_{19} nanoclusters via Density Functional Theory (DFT). First, we find in congruence with experimental results[4], that the clusters' energetics of hydrogen chemisorption (PCET) are exergonic up to $V_6O_{13}(OH)_6$, and that the cluster Bond Dissociation Free Energies (BDFEs) follow a slight decreasing trend with increasing degree of cluster hydrogen-chemisorption, similar to our previous group findings[5]. The DFT results were in strong agreement with experimental BDFE and BDFE-trend results, especially after accounting for configurational entropies.

We found that the BDFE trend is significantly influenced by the configurational entropies of the increasingly hydrogen-chemisorbed clusters. We also found that the configurational entropies are themselves impacted by the inclusion of near-iso-energetic spin states that increase in number with increasing hydrogen chemisorption. This work rationalizes a series of experimental results on PCET processes of POM clusters and enables us to understand how cluster electronic properties and ensemble effects give rise to individual PCET steps.

1. Chakraborty, S.; Schreiber, E.; Sanchez-Lievanos, K. R.; Tariq, M.; Brennessel, W. W.; Knowles, K. E.; Matson, E. M., Modelling local structural and electronic consequences of proton and hydrogen-atom uptake in VO2 with polyoxovanadate clusters. *Chemical Science* **2021**, 12 (38), 12744-12753.

2. Cooney, S. E.; Fertig, A. A.; Buisch, M. R.; Brennessel, W. W.; Matson, E. M., Coordination-induced bond weakening of water at the surface of an oxygen-deficient polyoxovanadate cluster. *Chemical Science* **2022**, 13 (43), 12726-12737.

3. Chakraborty, S.; Petel, B. E.; Schreiber, E.; Matson, Ellen M., Atomically precise vanadium-oxide clusters. *Nanoscale Advances* **2021**, 3 (5), 1293-1318.

4. Fertig, A. A.; Brennessel, W. W.; McKone, J. R.; Matson, E. M., Concerted Multiproton–Multielectron Transfer for the Reduction of O2 to H2O with a Polyoxovanadate Cluster. *Journal of the American Chemical Society* **2021**, 143 (38), 15756-15768.

5. Miu, E.V; McKone, J.R.; Mpourmpakis, G., The Sensitivity of Metal Oxide Electrocatalysis to Bulk Hydrogen Intercalation: Hydrogen Evolution on Tungsten Oxide. *Journal of the American Chemical Society* **2022**, 144 (14), 6420-6433.

84bn A DFT Analysis of Optimal Solvents for High-Throughput Processing of Imine-Linked Covalent Organic Frameworks

Emily Polo-Rankin, Alex Fatouros, Obioma Uche

Hydrogen-bonded lamellar solids have shown promise as candidates for next-generation extreme strength materials as they combine the advantages of traditional Kevlar and more recent graphene-like materials. Unfortunately, there has been limited success in finding cheap, non-volatile solvents for high-throughput solution-processing due to the above dual interactions. In this study, we utilize quantum mechanical calculations to understand the nature of the interactions between the 2-D polymer and different solvents (.e.g. water, trifluoroacetic acid etc.) in both vacuum and non-vacuum environments. Specifically, adsorption configurations, binding energies, and charge transfer between the polymer and solvents have been used to evaluate the strength of interactions and eventually, predict optimal solvents for these materials. Our results indicate that polymer-solvent interaction is primarily between the –OH group of the solvent and the imine-linkage of the COF with the strength of the interaction increasing with solvent acidity.

244f Investigation of Catalytic Activity of Fecunc Catalysts for Oxygen Reduction Reaction in Alkaline Medium with Gas Diffusion Electrode Half-Cell System

Jong Gyeong Kim, Hyeon-Seung Jung, Youngin Cho, Chanho Pak

Non-precious metal catalysts (NPMCs) are promising alternatives to expensive platinum catalysts for the cathode of the fuel cell. The lower overpotential of ORR and less corrosive environment at alkaline conditions made NPMCs more attractive [1]. To achieve efficient ORR catalysis, there are two factors to be considered. First, the development of high porosity of the catalyst for the formation and accessible active sites ($Fe\text{-}N_x$) [2]. The other is the binding energy between active sites and oxygen intermediates, which can be controlled by active site modification of coordination environment, heteroatom doping, additional $M\text{-}N_x$ sites, formation of metal nanoparticles, etc [3]. Following these strategies, the FeCuNC catalyst with superior ORR activity in an alkaline medium was prepared using SBA-15, a mesoporous silica template [4].

This study aims to investigate the catalytic behavior of FeCuNC at a wide potential range including kinetic and mass transport limiting regions for a profound understanding of ORR catalysis. First, the catalytic activities of FeCuNC were investigated using a rotating disk electrode (RDE) system. Although enhanced kinetics of FeCuNC (0.92 V @ 3 mA cm^{-2}) relative to FeNC (0.89 V @ 3 mA cm^{-2}) and commercial Pt/C (0.83 V @ 3 mA cm^{-2}) was confirmed at the high potential region, limitations of analyzing the mass-limiting region of the polarization curve still existed due to the low solubility (1.26×10^{-6} mol cm^{-3}) and the diffusion coefficient (1.93×10^{-5} cm^2 s^{-1}) of oxygen in 0.1 M KOH [5]. Therefore, a gas diffusion electrode (GDE) half-cell system was introduced to investigate the catalytic activity of FeCuNC in the mass-limiting region. GDE was used as a working electrode contacted with flow channels for oxygen gas supply. By controlling variables such as catalyst loading of GDE, ORR behavior could be investigated in a condition relatively similar to that of a real single cell rather than RDE. GDE half-cell could provide information about both the kinetics of the catalyst and the mass transport effect due to the pore structure more effectively. Analysis of the resistance component of GDE by electrochemical impedance spectroscopy enabled a more profound understanding of the cathode-specific resistance with NPMCs.

[1] N. Chen, Y. M. Lee, *Prog. Polymer Sci.*, **113** (2021) 101345.

[2] S. H. Lee, J. Kim, D. Y. Chung, J. M. Yoo, H. S. Lee, M. J. Kim, B. S. Mun, S. G. Kwon, Y. -E. Sung, T. Hyeon, *J. Am. Chem. Soc.*, **141** (2019) 2035.

[3] K. Wan, T. Chu, B. Li, P. Ming, C. Zhang, *Adv. Sci.*, (2023) 2203391.

[4] J. G. Kim, J. Cho, S. Han, H. Lee, E. Yuk, B. Bae, S. S. Jang, C. Pak, *J. Mater. Chem. A*, **10** (2022) 5361.

[5] B. B. Blizanac, P. N. Ross, and N. M. Marković, *J. Phys. Chem. B*, **110** (2006) 4735.

84bo Transport of Pseudomonas Aeruginosa in Native Respiratory Mucus

Riley Dickson, Zhijian He, Liheng Cai

Respiratory mucus is a viscoelastic hydrogel that serves as the first line of defense for the lungs. As we breathe, we inhale bacterial, viral, and environmental particulates into our lungs. Mucus entraps these particulates and is transported out of the lungs through coordinated ciliary beating of the bronchial epithelia, reminiscent of transporting people using an escalator, via the mechanism mucociliary clearance. The efficiency of mucociliary clearance becomes impaired when mucus concentration is altered. For instance, the concentration of airway mucus becomes abnormally high for patients with mucus-obstructive lung diseases like cystic fibrosis or chronic obstructive pulmonary disease, and the lungs become highly susceptible to bacterial infections. Although restoring the properties of mucus is critical to treatment of these diseases, this requires elucidating the roles of mucus as a physical hydrogel in defense.

In this work, I seek to develop a fuller understanding of the roles of native mucus in the context of mucociliary clearance. I focus on two aspects: (1) mucus-bacteria interactions and (2) mucus properties essential to efficient mucus clearance. I develop a method for harvesting native mucus with concentrations matching healthy and diseased airway mucus using well-differentiated human bronchia epithelial cell cultures. I quantify the transport of opportunistic pathogen Pseudomonas aeruginosa strain PA14, a human clinical isolate responsible for chronic lung infections, in this native mucus to elucidate the effects of mucus-bacteria interactions on bacteria motility. I find that PA exhibits a bimodal motility; in healthy mucus, motion is primarily linear, with a small proportion adopting a confined motion. As mucus concentration increases, PA14 exhibit an increasing propensity for confinement. With this work, I seek to unravel long-standing questions of human lung defense by understanding the mechanisms of bacterial transport through native mucus.

84bp QFDFT and QFConfsearcDFT

Laszlo Fusti-Molnar

Two new applications will be presented on this poster for the drug design community. The QFDFT application is a new, accurate and extremely efficient DFT program for high throughput and accurate ab-initio DFT calculations using modern dispersion corrections. The second QFConfSearchDFT application provides robust search in conformational space followed by QM semi-empirical geometry optimizations and very accurate ab-initio DFT energy calculations in order to establish accurate rankings and strain energies.

Developing Comprehensive Digital Twins of Research Laboratories: Introducing a Systems Engineering Approach to Lab Automation

Simon Rihm, Aleksandar Kondinski, Jiaru Bai, Markus Kraft

In order to tackle many of humanity's most pressing challenges, scientific discovery needs to be substantially accelerated. The automation of research activities plays a big role in this, from ubiquitous software tools to "self-driving laboratories". Recently, the idea of an "AI scientist" that can make Nobel-worthy discoveries has been introduced in this context [1]. We argue that the current platform-based "bottom-up" approaches are not sufficient to achieve and could even limit further development [2].

Therefore, we introduce a holistic lab automation framework as part of The World Avatar project, an all-encompassing digital twin based on dynamic knowledge graph. Its hierarchical and semantic structure allow for deep representation of knowledge across different domains and scales. We argue it necessary to further interoperability by widening the search and optimisation space to include managerial tasks in research labs as well as information on infrastructure and buildings. This way we can ensure cost effectiveness, improve reproducibility, and free up human resources for creative tasks. To achieve this, an unconventional "top-down" approach is applied to lab automation, taking a systems engineering perspective. This approach aims to integrate all aspects of lab work and its automation, contrasting the many isolated solutions available that – amongst others – increase the risk of manufacturer lock-in.

It enables the deployment of a multi-agent system in which humans are inherently embedded [1]. By considering the human-in-the-loop as well as the surrounding infrastructure, the definition of objectives can go beyond the optimisation of reaction conditions and include aspects of cost effectiveness, resource distribution and safety. A classical self-driving lab can be enhanced to derive its own goals, execute optimisation tasks in a distributed manner [2], consider its direct and indirect environment, and potentially integrate with complex multiscale models [3] to iteratively improve the digital twins' comprehensiveness and validity.

References

[1] Kitano, H. Nobel Turing Challenge: creating the engine for scientific discovery. *npj Syst. Biol. Appl.* **7**, 1–12 (2021).

[2] Bai, J. *et al.* From Platform to Knowledge Graph: Evolution of Laboratory Automation. *J. Am. Chem. Soc. Au* **2**, 292–309 (2022).

[3] Rihm, S.D. *et al.* Fully Automated Kinetic Models Extend our Understanding of Complex Reaction Mechanisms. *Chem. Ing. Tech.* (2023).

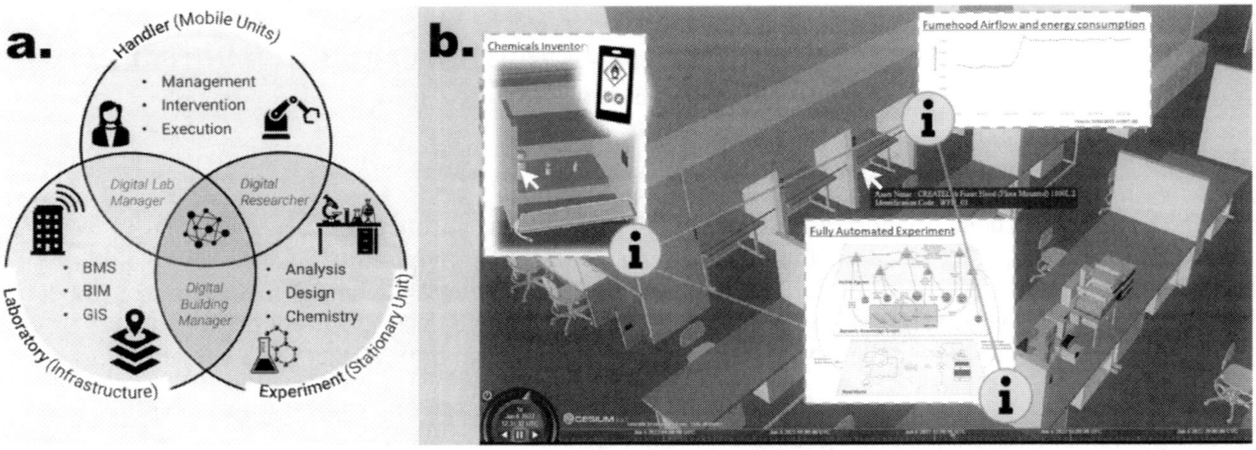

Figure: Overview of a holistic framework for research lab automation proposed. On the left side **(a)** a venn diagram illustrates the separation of functionalities and the specific potential for the creation of an "AI scientist" emerging at the respective interfaces. On the right side **(b)** the 3D visualisation aspect of a comprehensive digital twin for research laboratories is shown, overlaid with details on its functional aspects in three representative use cases.

Automating Rational Design of Polyhedral and Reticular Assemblies with Knowledge Engineering

Aleksandar Kondinski, Sebastian Mosbach, Jethro Akroyd, Markus Kraft

Knowledge Engineering (KE) has played a pivotal role in advancing chemistry since the 1960s, by simulating human expert decision-making processes and creating various expert systems.[1] Early applications, such as automated retrosynthetic analyses and mass spectrum elucidations, demonstrated KE's capacity for tackling complex problems and generating valuable hypotheses. Over the past two decades, KE has undergone significant advancements with the advent of the Semantic Web and knowledge graphs (KGs), fostering broader interoperability and machine-actionable knowledge. These developments have laid the foundation for rapid development of software agents able to traverse KG data, reason, and generate new knowledge, services, and discoveries in chemistry.

In this contribution, we present a KE-driven approach for automating the rational design of complex assemblies such as metal-organic polyhedra (MOPs)[2] and covalent organic frameworks (COFs).[3] Unlike the classical retrosynthetic analysis which can be carried out in a linear fashion, the design of these systems requires simultaneous consideration of chemical and spatial complementarity between multiple building units, making them challenging. Our approach leverages knowledge engineering and formalization in the form of ontologies, which are used to instantiate carefully curated data and develop agents that streamline the design process, navigate the chemical space, and generate candidate structures with explainable synthetic routes. Furthermore, we highlight our method's adaptability in tackling broader problems in reticular and supramolecular chemistry, fostering stronger connections between different domains of chemistry and involving communities.

[1] A. Kondinski, J. Bai, S. Mosbach, J. Akroyd, M. Kraft Acc. Chem. Res. 2023, 56, 2, 128–139.
[2] A. Kondinski, A. Menon, D. Nurkowski, F. Farazi, S. Mosbach, J. Akroyd, M. Kraft Automated Rational Design of Metal-Organic Polyhedra. J. Am. Chem. Soc. 2022, 144, 11713–11728.
[3] A. Kondinski et al. [manuscript in preparation].

AUTHOR INDEX

AIChE
120 Wall Street, FL 23
New York, NY 10005-4020

ISBN 978-1-7138-9321-9

CURRAN ASSOCIATES INC.
proceedings.com